Mathias Noe, Chistof Humpert, Wescley de Sousa, Stefan Fink,
Friedhelm Herzog, Stefan Huwer, Marvin Nießen, Sebastian Pfaller,
Stephan Pöhler, Tobias Rodler, André Schmid, Tabea Arndt

380 kV Superconducting Fault Current Limiter Feasibility Study

Karlsruher Schriftenreihe zur Supraleitung Band 037

HERAUSGEBER

Prof. Dr. Tabea Arndt
Prof. Dr. rer. nat. Bernhard Holzapfel
Prof. Dr. rer. nat. Sebastian Kempf
Prof. Dr.-Ing. Mathias Noe

Eine Übersicht aller bisher in dieser Schriftenreihe erschienenen Bände finden Sie am Ende des Buches.

380 kV Superconducting Fault Current Limiter Feasibility Study

by
Mathias Noe, Chistof Humpert, Wescley de Sousa,
Stefan Fink, Friedhelm Herzog, Stefan Huwer,
Marvin Nießen, Sebastian Pfaller, Stephan Pöhler,
Tobias Rodler, André Schmid, Tabea Arndt

Scientific
Publishing

TenneT

Technology Arts Sciences
TH Köln

KIT
Karlsruher Institut für Technologie

Impressum

KIT Scientific Publishing

Karlsruher Institut für Technologie (KIT)
KIT Scientific Publishing
Straße am Forum 2
D-76131 Karlsruhe

www.ksp.kit.edu

Print on Demand 2023 – Gedruckt auf FSC-zertifiziertem Papier

ISSN 1869-1765
ISBN 978-3-7315-1313-1
DOI 10.5445/KSP/1000161057

List of authors in alphabetical order

Tabea Arndt, KIT

Wescley de Sousa, KIT

Stefan Fink, KIT

Friedhelm Herzog, Messer

Christof Humpert, TH Köln

Stefan Huwer, Ingenieurbüro Huwer

Marvin Nießen, TH Köln

Mathias Noe, KIT, coordinator

Sebastian Pfaller, TenneT TSO GmbH

Stephan Pöhler, TenneT TSO GmbH

Tobias Rodler, TenneT TSO GmbH

André Schmid, TH Köln

Adresses of partners

Prof. Dr.-Ing. Mathias Noe Institute for Technical Physics Karlsruhe Institute of Technology Hermann-von-Helmholtz Platz 1 76344 Eggenstein-Leopoldshafen Tel.: +49 721 608 23500 Email: Mathias.noe@kit.edu	Prof. Dr.-Ing. Christof Humpert Institute of Electrical Power Engineering TH Köln - University of Applied Sciences Betzdorfer Straße 2 50679 Köln Tel.: +49 221 8275 2277 Email: christof.humpert@th-koeln.de
Hon.-Prof. Dr.-Ing. Stephan Pöhler Asset Management | Asset Technology Technical Consultant TenneT TSO GmbH Bernecker Straße 70 95448 Bayreuth Tel.: +49 921 50740 4608 Email: stephan.poehler@tennet.eu	Dr.-Ing. Friedhelm Herzog Messer SE & Co. KGaA Senior Manager Application Technology Industry Kleinewefersstraße 1 47803 Krefeld Tel.: +49 2151 7811 225 Email: Friedhelm.Herzog@messergroup.com
Dipl.-Ing. Stefan Huwer Ingenieurbüro Huwer Im Flürchen 29 67715 Geiselberg Tel.: +49 170 38 25 489 Email: Stefan@Huwer.de	

Contents

List of Figures

List of Tables

List of Abbreviations

AC	Alternating Current
AIS	Air Insulated Substation
CAPEX	Capital Expenditures
CB	Circuit Breaker
CLR	Current Limiting Reactor
DC	Direct Current
DMD	Diolen – Mylar - Diolen
EEC	Electrical Equivalent Circuit
FACTS	Flexible AC Transmission System
FCL	Fault Current Limiter
FEM	Finite Element Method
FRP	Fibre Reinforced Plastic
GM	Gifford MacMahon
GRP	Glass Fibre Reinforced Plastic
HTS	High-Temperature Superconductor
LI	Lightning Impulse
LIC	Chopped Lightning Impulse
LN_2	Liquid Nitrogen
OPEX	Operational Expenditures
RLC	Resistance, Inductance, Capacitance
RT	Room Temperature
SFCL	Superconducting Fault Current Limiter
SI	Switching Impulse
TOTEX	Total Expenditure (CAPEX and OPEX)
TP	Triple Point
WACC	Weighted Average Cost of Capital

List of Symbols

A Cross section

A_{HTS} Cross section of HTS tapes

C Capacitance

c Light velocity

c_p Specific heat capacity

$c_{p, HTS}$ Specific heat capacity of HTS tapes

d Thickness, diameter

d_i Inner diameter

d_o Outer diameter

d_{tt} Distance turn to turn

d_{HTS} Density of HTS tapes

E Electric field

I Current

i_{max} Maximum limited current

i_p Prospective peak short-circuit current

i_r Rated current

I_c Critical current of the superconductor

I_k'' Prospective short-circuit current (RMS)

I_n Nominal current

j_c Critical current density

k Coupling factor

l Length

L Inductance

l_{HTS} Length HTS tapes

L_Q Network inductance

L_s Network inductance

n Exponential factor for E-J curve of HTS

P Loss

p_l Number of HTS tapes in parallel

t Time

T Temperature

T_C Critical temperature of a superconductor

t_a Reaction time of fault current limiter

t_d Fault duration

t_r Recovery time of fault current limiter

U	Voltage
U_{m}	Maximum voltage of equipment
U_{n}	Nominal voltage
U_{LI}	Lightning impulse voltage
U_{LIC}	Chopped lightning impulse voltage
U_{SI}	Switching impulse voltage
V	Volume
V_{HTS}	Volume of HTS tapes
$\dot{Q}$	Heat flow
r	Radius
R	Resistance
R_{HTS}	Superconductor Resistance
R_{P}	Parallel resistance
R_{Q}	Network resistance
X	Reactance
Z	Impedance
Z_{Sh}	Shunt impedance
ε_{r}	Permittivity
λ	Thermal conductivity
η	Schwaiger factor
ρ_{HTS}	Resistivity of HTS tapes
μ_0	Vacuum permeability
τ	Propagation time of travelling wave

Acknowledgement

We are very grateful to TenneT TSO GmbH for supporting this study. The authors would like to acknowledge the support of Dr. Wolfgang Reiser and Dr. Joachim Bock for valuable discussions about the manufacturing of HTS fault current limiting modules and the feedback from several HTS manufacturers about the performance and cost of HTS tapes. We gratefully acknowledge the proof reading of Pauline Leys and the critical comments from Prof. Steffen Elschner. Both feedbacks improved the readability of this study a lot.

1 Executive Summary

The main motivation for this study is, that short-circuit currents increase at the 380 kV level and that there is no conventional component that can reduce the short-circuit current effectively within the first half-wave of the short-circuit current while having no impedance at normal operation.

In this study the technical and economical feasibility of a 380 kV, 5 kA resistive type superconducting fault current limiter (SFCL) was investigated for a bus coupling (case A) and bus section coupling (case B) location. Since this type of SFCL had only been developed up to 220 kV, this is beyond the state of the art and emphasis was therefore given on various high voltage design aspects. The main findings of this study are:

- Assuming a critical current of 700 A per tape at 77 K and self field, 235/209 km (case A/case B) of 10-12 mm wide HTS tapes are needed for a three phase 380 kV, 5 kA resistive type SFCL. Two tapes need to be connected in parallel as so-called „twin tapes“ in order to enable a compact design. The material should be available from several manufacturers within a reasonable development time.
- The HTS tapes are arranged in bifilar coils in so-called HTS modules. The complete current limiting element consists of
    - 85 HTS modules with an outer diameter of 1100 mm for case A and
    - 93 HTS modules with an outer diameter of 1000 mm for case B.
- The superconductor is cooled in liquid nitrogen at an operating temperature of 77 K and a pressure of 5 bar. Liquid nitrogen serves as a coolant and as electrical insulation between the limiting element at high voltage and the cryostat at ground voltage.
- In each phase, a current limiting reactor with an inductance of 72.9/39.5 mH (case A/case B) is connected in parallel to the superconducting limiting element. This increases the reliability of the system and reduces the amount of HTS tapes required.
- The maximum electric field inside the limiter between high voltage parts and the grounded cryostat wall is estimated to be 3.8 kV/mm. Due to a higher field for a sphere in comparison to a cylinder geometry, the end flanges of the cryostat define the minimum diameter of the cryostat.
- The series connection of the HTS modules leads to a non-linear voltage distribution along the modules at short-circuit. An RLC network model was developed and the voltage distribution along the HTS modules was calculated. The asymmetric voltage distribution along the HTS modules seems manageable and the electric field strength in the winding insulation seems feasible as well.
- The measurement and simulation of travelling waves within a bifilar coil showed clearly that a transient voltage pulse propagates in radial and circumferential direction. This leads to an increased voltage stress within the winding insulation. In this simulation, an LIC pulse could lead to a maximum voltage in the winding insulation of 160 kV. This voltage oscillates with about 8 MHz, which leads to a pulse length of 60 ns only.
- For one phase, the HTS modules can fit in both cases A and B in a cryostat with a length of 13.6 m and a diameter of 3.6 m. The weight of one phase of the SFCL is 142 tons, thereof 94 tons of liquid nitrogen.

- The three phase system consists of three SFCLs, three current limiting reactors and the cooling system with the following dimensions:
  - Three phase SFCL – 16.6 m x 13.6 m x 10.1 m (LxWxH)
  - Three current limiting reactors – 10.3 m x 9.4 m x 8.3 m (LxWxH)
  - Open cooling system – one standard 20 ft container plus a larger LN_2 reservoir
  - Closed cooling system – two standard 20 ft container plus a smaller LN_2 reservoir
- A reliable cooling system can be realized with an open or closed cooling system. In an open system, a liquid nitrogen reservoir needs a regular refill, while a closed cooling system provides the cooling power with cryocoolers. The maximum cooling power at rated current at 77 K is estimated to 17.6/15.9 kW (case A/case B). Due to the impact of the AC losses of the HTS tapes, this is much higher than the cooling power at medium currents.
- The investment cost for a three phase 380 kV, 5 kA SFCL ranges between 19-37.5/18.2-53.1 Mio. € (case A/case B). The HTS material cost has the highest cost share with more than 40 % and the highest uncertainty. The operation cost depends strongly on the load of the SFCL. For a load factor of 0.7, the operation cost per year estimates to 315/289 k€. This rises to 695/628 k€ (case A/case B) per year for a load factor of one.

These main findings encourage us to the final statement, that the development of a 380 kV, 5 kA resistive SFCL is feasible. No technical showstoppers and huge obstacles were identified.

As a next step it is recommendable to develop the key components of a 380 kV, 5 kA resistive SFCL first. This is a 380 kV cryogenic bushing and the HTS current limiting module. Based on the results of this first two-year R&D project, a single-phase resistive type 380 kV, 5 kA SFCL can be developed within three years.

2 Introduction

2.1 Motivation

In electrical power systems, the limitation of short-circuit currents plays an important role in protecting the existing equipment. All components installed in the electrical network must be able to withstand the thermal and mechanical stresses in the event of a short circuit without damage or degradation. This requires adequate dimensioning of the components. To limit short-circuit currents, there is a wide range of network and equipment-related measures, which are summarized in Fig. 2.1.1. The classic network engineering measures to reduce short-circuit currents include the formation of subnetworks, the introduction of a higher voltage level to generally reduce currents as power increases, and busbar separation.

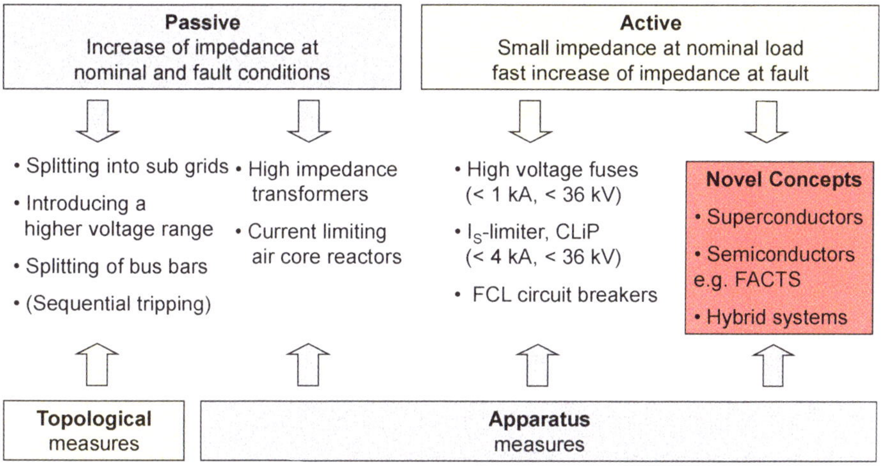

Figure 2.1.1: Conventional and novel measures for limiting short-circuit currents including superconducting fault current limiters [after CIGRE2012]

Today, the common factors in all measures is that they increase the network impedance in case of a fault as well as in normal operation. This is also true for the most commonly applied equipment-related measures today, such as transformer impedance or the use of additional reactors. The ideal solution would be an active current limiter that only increases the impedance instantaneously when a fault occurs and has a low or negligible impedance during normal operation. Up to medium voltage, HV fuses or so-called Is-limiters are in use [ABB2017], but their insert needs replacement after each trip, and they are not intrinsically safe. This prevents a wide application despite a high technical reliability.

Among the novel concepts that allow a rapid increase in impedance in the event of a fault with low impedance in normal operation, superconducting current limiters are currently the most advanced. To date, superconducting current limiters have been successfully tested in grids up to 220 kV [Moyzikh2021], and the first manufacturers from Russia and Korea have begun to offer commercial solutions. Figure 2.1.2 shows the basic current limiting process with the important parameters. In normal operation, superconducting current limiters have very low or negligible impedance. As soon as the first current rise after the short circuit occurs, the prospective short circuit current i_p, without external trigger, is reduced to a maximum limited current i_{max}. During the fault duration t_d, the current is reliably and intrinsically limited. Normally, the fault is cleared by the network protection after the fault duration. After a recovery time t_r, which can also be zero depending on the requirement, the limiter is usually automatically available for recurring normal operation.

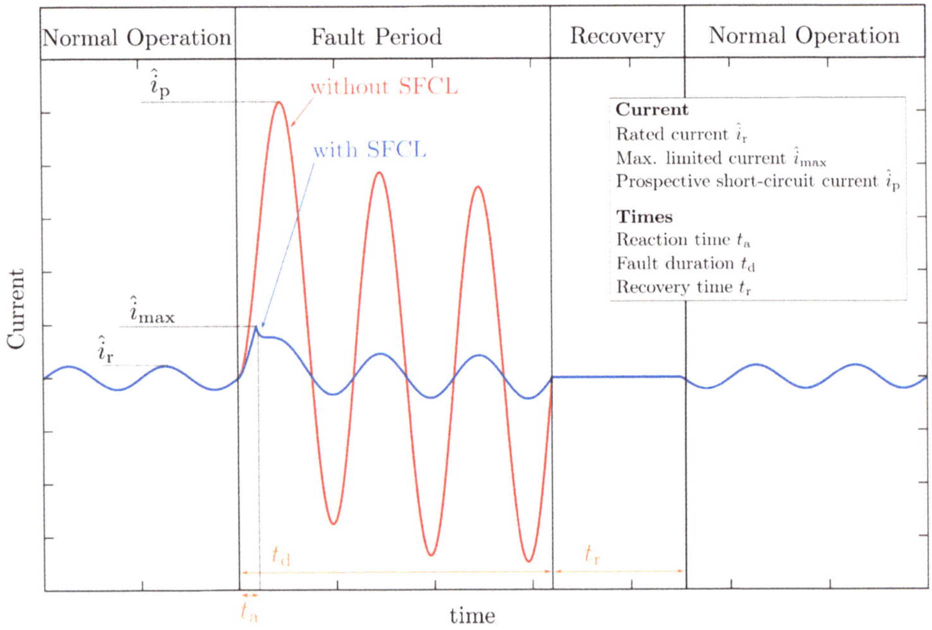

Figure 2.1.2: Typical current limiting process with important parameters

Superconducting current limiters have the characteristics listed below, which are comparable to those of an ideal current limiter.

- Independent, fast current limiting in the first current rise, no need of fault current detection
- Negligible or low impedance during normal operation
- Fast and automatic (without maintenance) recovery after tripping
- Intrinsically safe current limiting even if the superconductor is defective
- Possibility of application in high voltage

The combination of these properties of superconducting current limiters cannot achieved by conventional current limiting measures. Thus, superconducting current limiters have a unique selling point in comparison to other measures for limiting short-circuit currents.

2.2 Requirements

Table 2.2.1 summarizes the main requirements for the 380 kV resistive type superconducting fault current limiter. In this study, two similar applications were investigated. They differ only in the maximum limited current with a limiter. In the bus section coupling case B the maximum limited current is 30 kA (rms) whereas in case A this value is only 19 kA (rms). This results in a higher current limiting impedance for case A in comparison to case B. It is assumed that the nominal current is also the maximum current that flows through the limiter in normal operation.

Table 2.2.1: Main requirements for 380 kV fault current limiter

	[A] AIS Bus Coupler	**[B] Bus Section Coupling**
Nominal voltage, U_n	380 kV	380 kV
Temporary highest voltage for equipment	440 kV	440 kV
Nominal current, I_n	5.0 kA	5.0 kA
Max. short-circuit current without limiter, I_k''	63 kA	63 kA
Max. limited current with FCL, $I''_{k,lim}$	19 kA	30 kA
Fault duration t_d	0.25 s	0.25 s

At present, no international standard exists for testing superconducting fault current limiters. An international IEEE working group released in 2015 a guide for testing fault current limiters [IEEE2015]. In this guide detailed test procedures are given and for maximum test voltages references to other standards are made. For this study, the maximum voltages that are needed for a reliable high voltage design are based on the IEC standard IEC60076-3 for power transformers. The relevant test voltages are summarized in Table 2.2.2.

Table 2.2.2: High voltage requirements for 380 kV SFCL in accordance with IEC 60076-3

Description	Symbol	Data
Highest AC voltage for equipment	U_m	420 kV (rms)
Lightning impulse voltage	U_{LI}	1425 kV (peak)
Chopped lightning impulse voltage	U_{LIC}	1570 kV (peak)
Switching impulse voltage	U_{SI}	1175 kV (peak)
AC withstand voltage	$U_{AC\text{phase to earth}}$	630 kV (rms)

2.3 Resistive Type Superconducting Fault Current Limiter

Figure 2.3.1 shows the single-pole equivalent electrical circuit of the resistive superconducting fault current limiter in a simple electrical circuit. At currents up to the rated current, the high-temperature superconducting material is in the superconducting state and thus has no measurable resistance R_{HTS}. Due to the strong increase of the current in case of a fault beyond the critical current I_c[1] of the superconductor, an electrical resistance develops automatically (intrinsic to the material) and immediately, which causes the current limitation. As a result, the superconductor usually heats up to well above the critical temperature T_c. Therefore, the superconductor usually has to be disconnected after a certain time to recover to the initial state of stable operating conditions for normal operation.

It is possible to connect a shunt impedance $\underline{Z}_{Sh}$ in parallel to the high-temperature superconducting material. This can be a current limiting reactor (CLR) or any other impedance.

[1] The critical current I_c is defined as the current where the electric field reaches 1 µV/cm. At this value the resistivity of the superconductor rises sharply and the superconductor becomes normal conducting.

With this impedance, it is possible to adjust the current in the superconductor during limitation and as a consequence the heating of the superconductor. In addition, the reliability increases in case of damage of the superconductor.

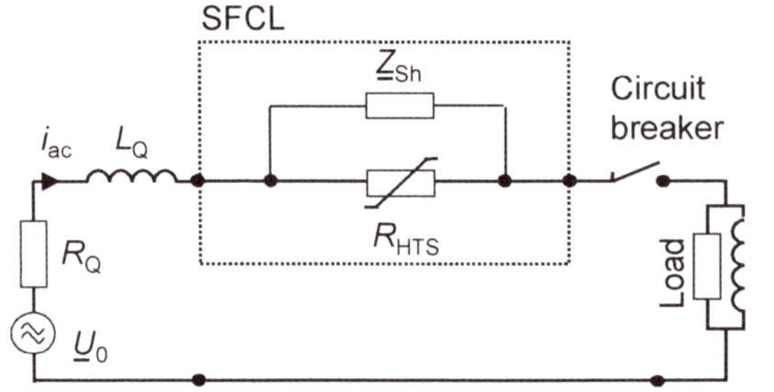

Figure 2.3.1: Electrical circuit of a resistive type superconducting fault current limiter

2.4 Methodology

Based on the main requirements given in chapter 2.2 and the basic function of a resistive type superconducting fault current limiter the first task is to estimate the amount of superconducting material. This is done by taking into account the state-of-the-art material parameters from various manufacturers and a maximum temperature calculation of the superconducting material during the fault duration t_d.

Secondly, based on these results, a reasonable current limiting module is designed for both applications. The current limiting module design already takes into account reasonable dimensions and AC losses of the superconductor. With these results a preferred current limiting module design is proposed and a detailed high voltage design is carried out for the preference design. This includes FEM calculations to estimate the maximum electrical field for critical areas, the investigation of travelling waves inside the limiter and the non-linear voltage distribution due to capacitive coupling effects. All this results in a design of one phase of a current limiter with detailed dimensions. With these data, the heat input to the coolant is estimated and the cooling system is conceptually designed. The conceptual design ends with a summary of the main parameters and a setup of the complete system.

The investigation of the economic feasibility in chapter 7 is based on this detailed design and the present worth method is used for this with a detailed analysis of investment and operation cost.

The description of the possible failures and their consequences in chapter 5 is based on the experience of the authors because so far, no detailed failure mode and effect analysis exists from SFCL manufacturers.

The chapter on SFCL tests in chapter 6 is also based on the experience of the authors that participated in several international CIGRE working groups to summarize the status of SFCL development and to develop a test recommendation guide.

3 High-Temperature Superconducting Material

3.1 Overview on Manufacturers and Availabilities

Currently, there are a few manufacturers of high-temperature superconductors worldwide that can supply material for superconducting current limiters in good quality and quantity and in a reasonable time. These are summarized in Table 3.1.1 and include SuperPower in the USA, Sunam in Korea, Shanghai Superconductors Technologies in China, S-Innovation in Russia and Theva in Germany. The material is the so-called 2nd generation high-temperature superconductor based on Re-Ba-Cu-O compounds. The abbreviation Re stands for rare earth, as different materials are used depending on the manufacturer.

Table 3.1.1: List of manufacturers of HTS tape material for use in fault current limiters

Company	Country	Weblink
SuperPower	US	www.superpower-inc.com
Shanghai Superconductors Technologies	China	www.shsctec.com
Sunam	Korea	www.i-sunam.com/wp/
Theva	Germany	www.theva.com
S-Innovations	Russia/Japan	www.eng.s-innovations.ru

3.2 Main Characteristics of available High-Temperature Superconducting Material for Fault Current Limiters

The conductor has the shape of a flat tape with a structure consisting of several layers. As an example, Fig. 3.2.1 shows the layered structure of Theva's superconductors.

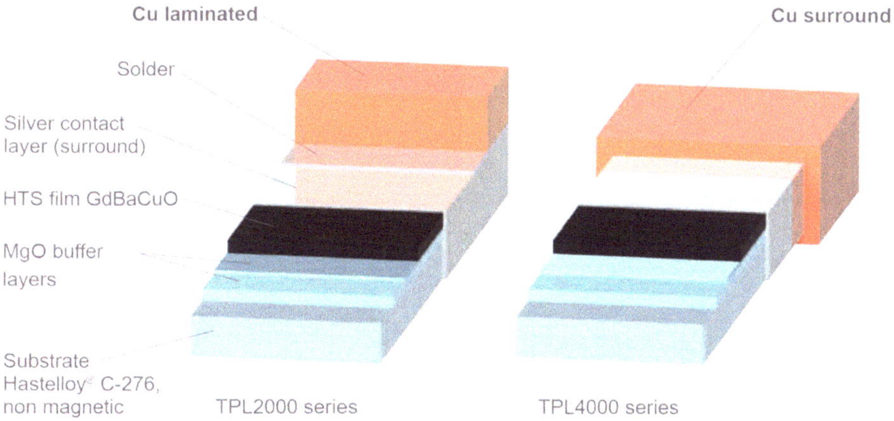

Figure 3.2.1: Layered structure of HTS tape from Theva Company [pictures from www.theva.com]

A substrate, often made of a Ni alloy with a thickness between 30 and 100 µm, serves as the basis. For most manufacturers, this is followed by several intermediate layers which serve as insulation between the substrate and the superconductor and give the superconductor a texture through their orientation. Only when the superconductor layers have a quasi monocrystalline pronounced bi-axial texture a significant current can flow. The superconductor layer is only

1-3 µm thick and can be applied to the interlayer by various physical or chemical processes [Noe2021]. In all manufacturing processes, the conductor is then surrounded by a silver layer of about 1 µm thickness. This protects the superconducting layer, which reacts strongly with water, acts as an electrical stabilizer and enables good electrical contacts of the ceramic superconducting materials.

The tape conductors are available with a width between 2 and 40 mm and with a total thickness between 40 and 100 µm. Single piece lengths of more than 100 m can be produced reliably and are thus sufficiently long for use in superconducting current limiters. The current density of superconductors is highly dependent on temperature, magnetic field and its direction. The conductors of the above-mentioned manufacturers currently have a critical current I_c in the intrinsic field at 77 K of about 400-800 A per centimeter of width. In the last 10 years, there has been more than a doubling of the critical current density and in short lengths well over 1000 A/cm have already been achieved at 77 K and intrinsic field. Thus, through continuous improvements in manufacturing processes, further increases in current-carrying capacity are foreseeable. For resistive superconducting current limiters, tapes with a width of 10 or 12 mm are usually used. All manufacturers offer various stabilizing layers of copper or other materials contacted on the silver layer.

3.3 Description of typical Quench Properties

The main function of a resistive type superconducting fault current limiter is given by the non-linear current-voltage curve that is shown in Fig. 3.3.1.

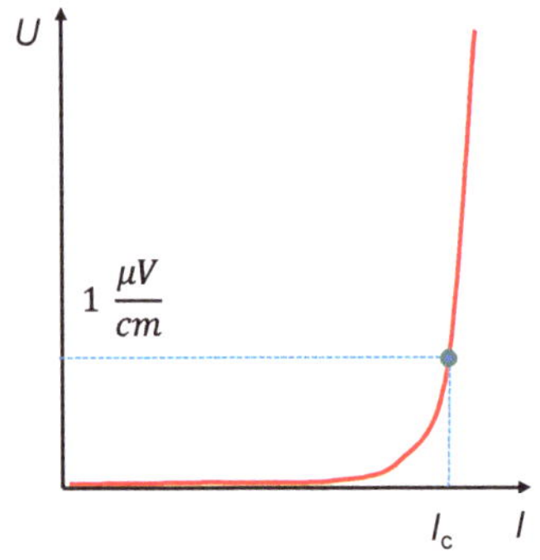

Figure 3.3.1: Typical current-voltage curve of a superconductor

A superconductor is characterized by its critical current I_c or critical current density j_c that is usually defined by the 1 µV/cm criteria. The drastic increase of the electrical field above the critical current density is in general parametrized by the following power law

$$E = E_c \left(\frac{j}{j_c}\right)^n \qquad (3.1)$$

with the exponent n in a typical range between 10 and 50 for HTS material, j_c as the critical current density of the HTS material and E_c as a constant value of 1 µV/cm.

At short circuit, the current exceeds the critical current and a resistance develops automatically that limits the current. Consequently, the superconductor is heated by the ohmic losses. A typical quench behaviour is shown in Fig. 3.3.2. Quench is the transition from the low-ohmic superconducting state to the high-ohmic normalconducting state of the superconductor.

Even for currents as low as two times the critical current, the short-circuit current is limited within the first half-wave of the current after approximately 4 ms. When the limitation starts, the resistance develops and the temperature rises. The higher the peak short-circuit current, the faster the limitation and the faster the rise in resistance and temperature. The residual current after the first limitation can be adjusted with an impedance in parallel to the superconductor. It is characteristic for resistive type SFCLs that the first sharp peak of the limited current is nearly independent of the peak short-circuit current without the limiter.

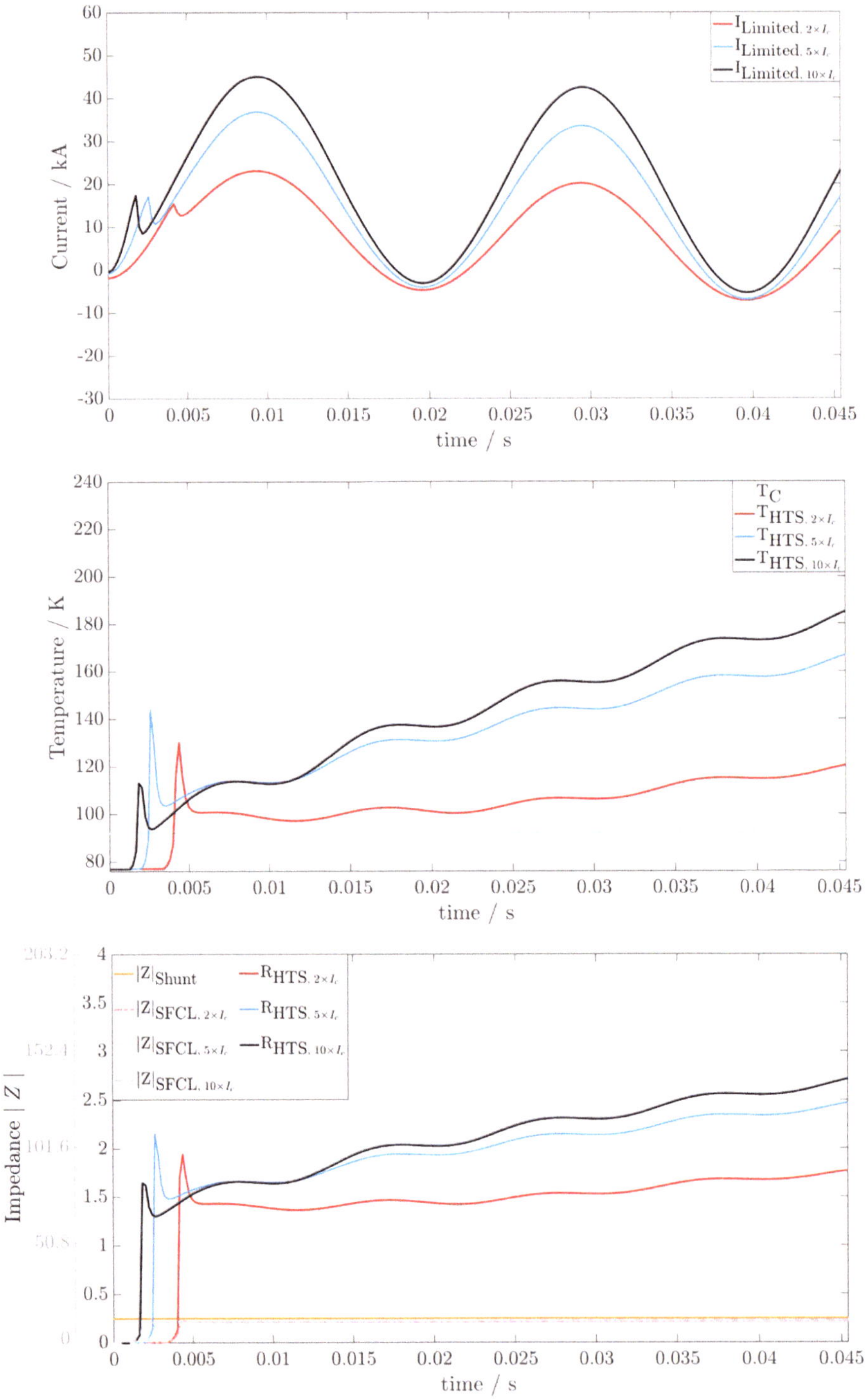

Figure 3.3.2: Current (top), temperature (middle) and resistance (bottom) curves for different peak short-circuit currents of 2 x , 5 x and 10 x I_c

4 Conceptual Design of a Resistive Type 380 kV, 5 kA Limiter

4.1 Estimation of Amount of Superconductor

The number of HTS tapes in parallel depends on the critical current of the tape. In general, the sum of the critical current of the HTS tapes in parallel must exceed the maximum continuous current in normal operation. This study assumes that the nominal current is equal to the maximum continuous current at normal operation. Table 4.1.1 shows the number of tapes in parallel for typical critical currents.

Table 4.1.1: Data of minimum number of tapes in parallel with different critical current

Critical current of one tape	**600 A**	**700 A**
Nominal current (rms)	5000 A	5000 A
Nominal current (peak)	7071 A	7071 A
Number of tapes in parallel	12	11
Total critical current	7200 A	7700 A
Ratio of I_{max} to I_c	0.982	0.918

The influence of the magnetic field on the critical current is neglected in the following. Since in the bifilar tape arrangement the magnetic fields are essentially compensated, this is justified. To calculate the required amount of superconductor, two important boundary conditions must be observed. Firstly, a maximum temperature of the superconductor must not be exceeded during the entire fault duration. Based on empirical values, a maximum temperature of the superconductor of 360 K is assumed here. With an operating temperature of 77 K in liquid nitrogen, the superconductor must then not heat up by more than 283 K during the entire limiting process. Equation 4.1.1 describes the adiabatic heating process of the superconductor during the limiting process. Within the first current half-wave this equation gives a good approximation of the heating of the superconductor.

$$c_{\mathrm{p,HTS}}\, d_{\mathrm{HTS}}\, V_{\mathrm{HTS}}\, \Delta T = \int_{t1}^{t2} \frac{u^2(t)}{R_{HTS}}\, dt \qquad (4.1.1)$$

where $c_{p,HTS}$ is the mean specific heat capacity, d_{HTS} is the density, V_{HTS} is the volume and R_{HTS} is the total resistance of the HTS tapes during limitation. Substituting the volume and the resistance, we obtain Eq. 4.1.2.

$$c_{\mathrm{p,HTS}}\, d_{\mathrm{HTS}}\, l_{\mathrm{HTS}}\, \cancel{A_{HTS}}\, \Delta T = \int_{t1}^{t2} \frac{u^2(t)\, \cancel{A_{HTS}}}{\rho_{\mathrm{HTS}}\, l_{\mathrm{HTS}}}\, dt \qquad (4.1.2)$$

where ρ_{HTS} is the mean resistivity, l_{HTS} is the length and A_{HTS} is the cross-section of the HTS tapes. Thus, the key parameter at a given voltage and material to limit the temperature rise is the superconductor length l_{HTS}, which depends on the following:

- The amount of superconductor increases nearly proportionally with the rated current as the number of tapes in parallel increase with rated current

- The superconductor length decreases proportionally with an increasing critical current I_c of the HTS tape because less tapes need to be connected in parallel.
- The superconductor length increases proportionally with the rated voltage as shown in Eq. 4.1.2.
- The superconductor length increases with the square root of the fault duration, as Eq. 4.1.2 shows. Thus, doubling the fault duration would mean 1.41 times more superconductor length.
- The superconductor length further depends on the material parameters of the specific heat capacity $c_{p,HTS}$, the density of the material d_{HTS} and the specific resistance ρ_{HTS}. However, these are specified by the material manufacturers and similar materials are used here, so this influence is rather small.

The cross section of the superconductor A_{HTS} has no influence on the temperature increase and the maximum temperature after current limiting can only be reduced by increasing the superconducting length. The maximum temperature decreases inversely proportional to the square of the superconductor length. The expected temperature increases and thus the length needed to maintain the HTS below 360 K can be simulated in a straightforward way according to Eq. 4.1.2 on the basis of a simplified short-circuit circuit as shown in Fig. 4.1.1.

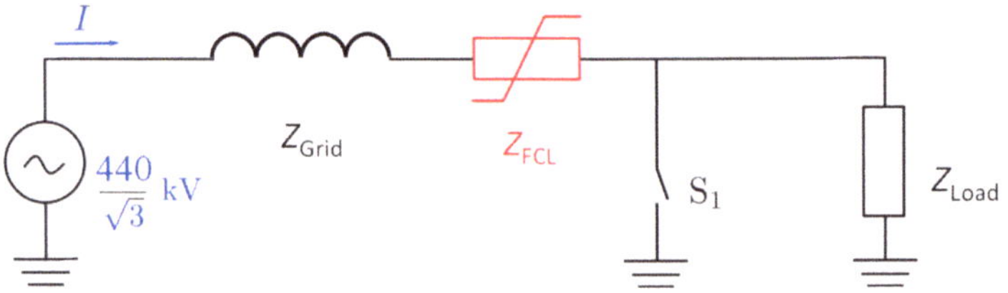

Figure 4.1.1: Simplified electrical circuit to calculate the temperature increase of superconductor and the maximum limited current during fault duration

A thorough analysis has to include the temperature dependence of the material parameters which are well known and summarized in Annex 12.1 and 12.2. In addition, the simulation includes the cooling of the superconductor since adiabatic heating is only given within the first current half-wave. With the long short-circuit times present here, the assumption of adiabatic heating would lead to much higher superconductor temperatures and consequently to a greater superconductor length. The heat transfer from the superconductor to the surrounding liquid nitrogen is shown in Annex 12.2. These curves assume pool-boiling conditions at 77 K. Heat transfer curves for pressurized liquid nitrogen are not yet widely investigated and data is shown in Fig. 4.1.2. At higher pressures a smaller temperature difference is needed to get the same heat transfer and the point of maximum heat transfer is shifted first to higher values and smaller temperature differences. This study uses the values for one bar and considers in addition the worst case without cooling. The difference in temperature increase or amount of material between the cooling at one bar and no cooling is calculated and shown in the following step. As demonstrated in the following chapter the current limitation is only slightly affected by the heat transfer parameter. The main difference is in the maximum temperature of the superconductor that is in worst-case conditions without cooling approximately 40 K higher than with cooling at

1 bar. Since the design of the limiter allows a good cooling of the superconductor even with 5 bar, the difference between 5 bar and 1 bar cooling is not considered to be that severe.

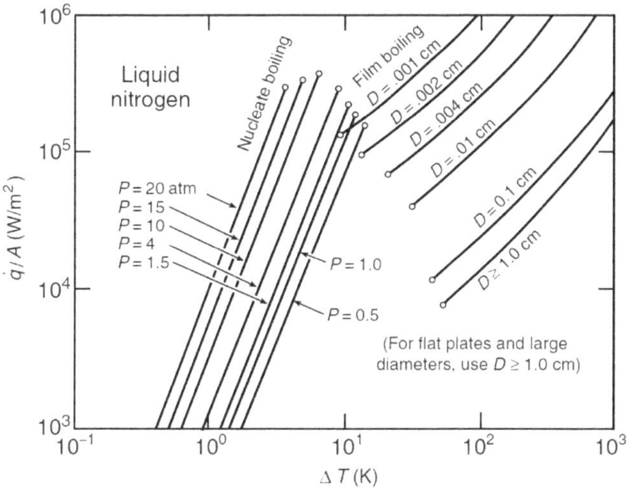

Figure 4.1.2: Heat transfer curves for a metal surface to liquid nitrogen and different pressures [Ekin2006]

The second important boundary condition is that during the limiting process the limited current must not exceed the maximum allowable current (see Table 2.2.1). This value can be influenced by two parameters. A longer superconductor length increases the superconductor resistance and thus lowers the limited current. The superconductor as a ceramic material becomes highly resistive in the normal line. An additional low-resistance copper layer on the superconductor then greatly affects the current-limiting resistance. A larger copper cross-section thus increases the limited current. In any case, a resistor or inductor can be connected in parallel with the superconductor to influence the limited current.

The required superconductor quantity is thus determined by the length at which both boundary conditions (max. temperature and max. permissible current) are met. Since a coating of the superconductor with copper has an influence on the limited current and the re-cooling at zero current, a superconductor with a copper layer of 10 µm was also investigated for comparison purposes. In principle, a coating with a high-resistance material would be advantageous for a resistive current limiter, but this is not yet available in a standardized form by the manufacturers and is therefore not considered in this investigation.

The results of the calculation for the required length of the HTS tapes are summarized in Table 4.1.2 (data of HTS tapes given in Annex 12.1). It can be seen that with a shunt impedance in parallel to the superconducting tapes, case A requires approximately 10 % more HTS tapes than case B. This is explained by the higher impedance necessary to achieve a lower limited current. In the course of this study the variant without shunt is no longer followed because it is not preferred by the utility and the tape variant with a critical current of 600 A is not further investigated because it needs considerably more tape.

Table 4.1.2: Summary of superconductor HTS tape length (further preferences in bold blue)

	Without Shunt Impedance		With Shunt Impedance	
	Case A	**Case B**	**Case A**	**Case B**
Critical current (tapes) / A	600/700	600/700	600/**700**	600/**700**
Total critical current / A	8400/8400	8400/8400	8400/**8400**	8400/**8400**
Number of parallel tapes	14/12	14/12	14/**12**	14/**12**
Length one tape / km	9.58/7.6	9.6/7.6	9.56/**6.5**	7.4/**5.8**
Length per phase / km	134.1/91	134.4/91.2	135.1/**78.4**	103.6/**69.7**
Total length (3 Ph.) / km	402.2/272.9	403.2/273.6	405.3/**235.2**	310.8/**209.1**

4.2 Current Limitation

The simulation of the current limitation, the temperature of the HTS material and the impedance of the fault current limiter is shown for case A in Fig. 4.2.1 and for case B in Fig. 4.2.2. Material data from Table 4.1.2 is used with the tape parameters according to Annex 12.1 and 12.2. Figure 4.2.1 shows that the maximum current is limited below the specified value of 26.87 kA ($\sqrt{2} \cdot 19$ kA) and that the maximum temperature of the HTS material is kept below 360 K at the end of the fault duration. Even in worst-case conditions of no cooling of the HTS material at fault duration, the maximum temperature remains below 400 K. For the heat transfer to the liquid nitrogen pool boiling conditions are assumed since reliable data under pressure is rarely available. Shortly after the quench of the superconductor, the resistance rises to a much higher value of the shunt in parallel. Therefore, the limiting impedance is dominated by the shunt impedance and is predominantly inductive. A very similar behaviour is seen in case B. The main difference in case B is the much higher limited current, mainly due to a different shunt impedance.

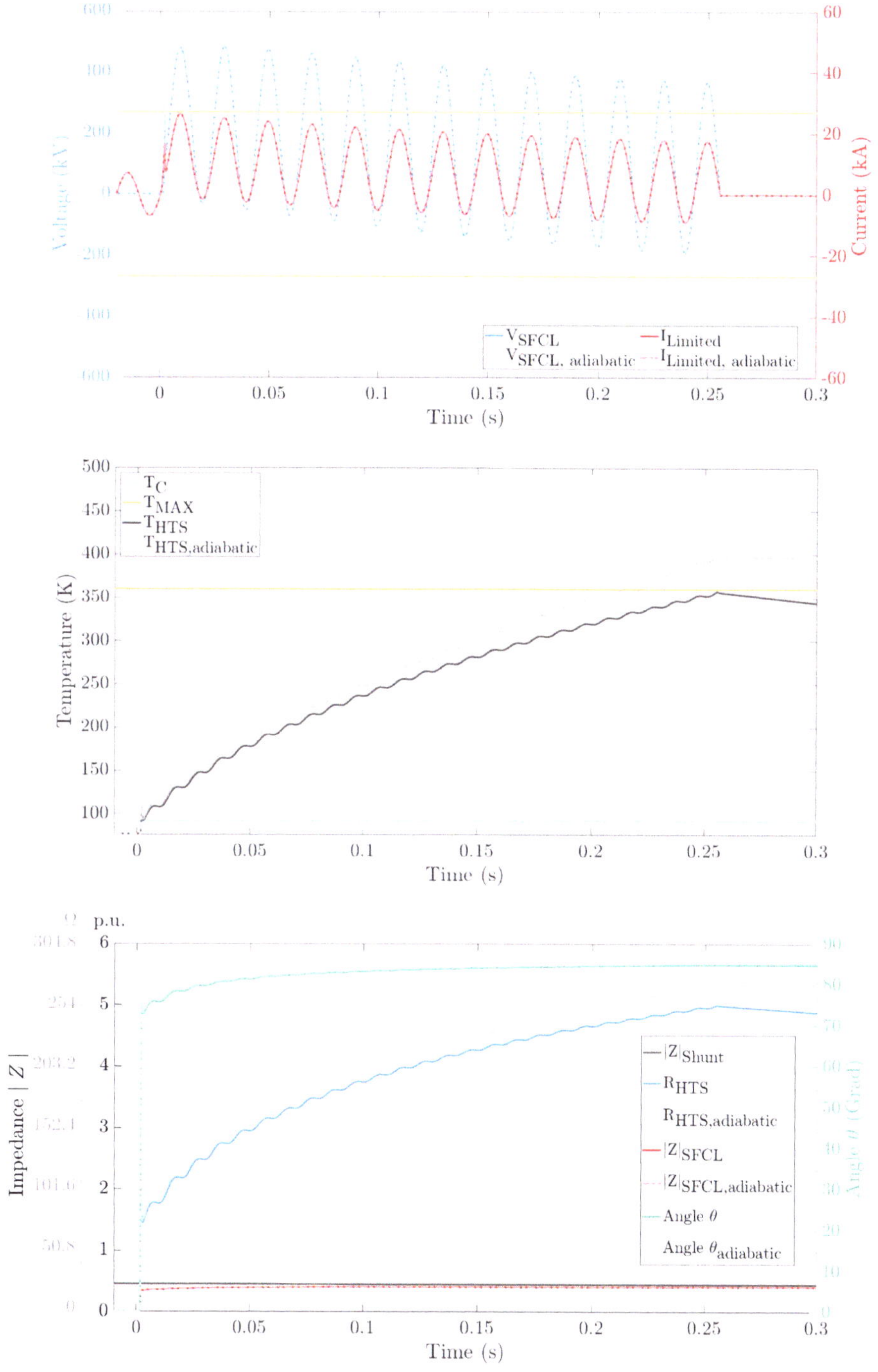

Figure 4.2.1: Current-Voltage curve (above), temperature curve of the HTS material (middle) and impedance curve of the limiter for Case A (Current and voltage are similar for adiabatic and non-adiabatic simulation)

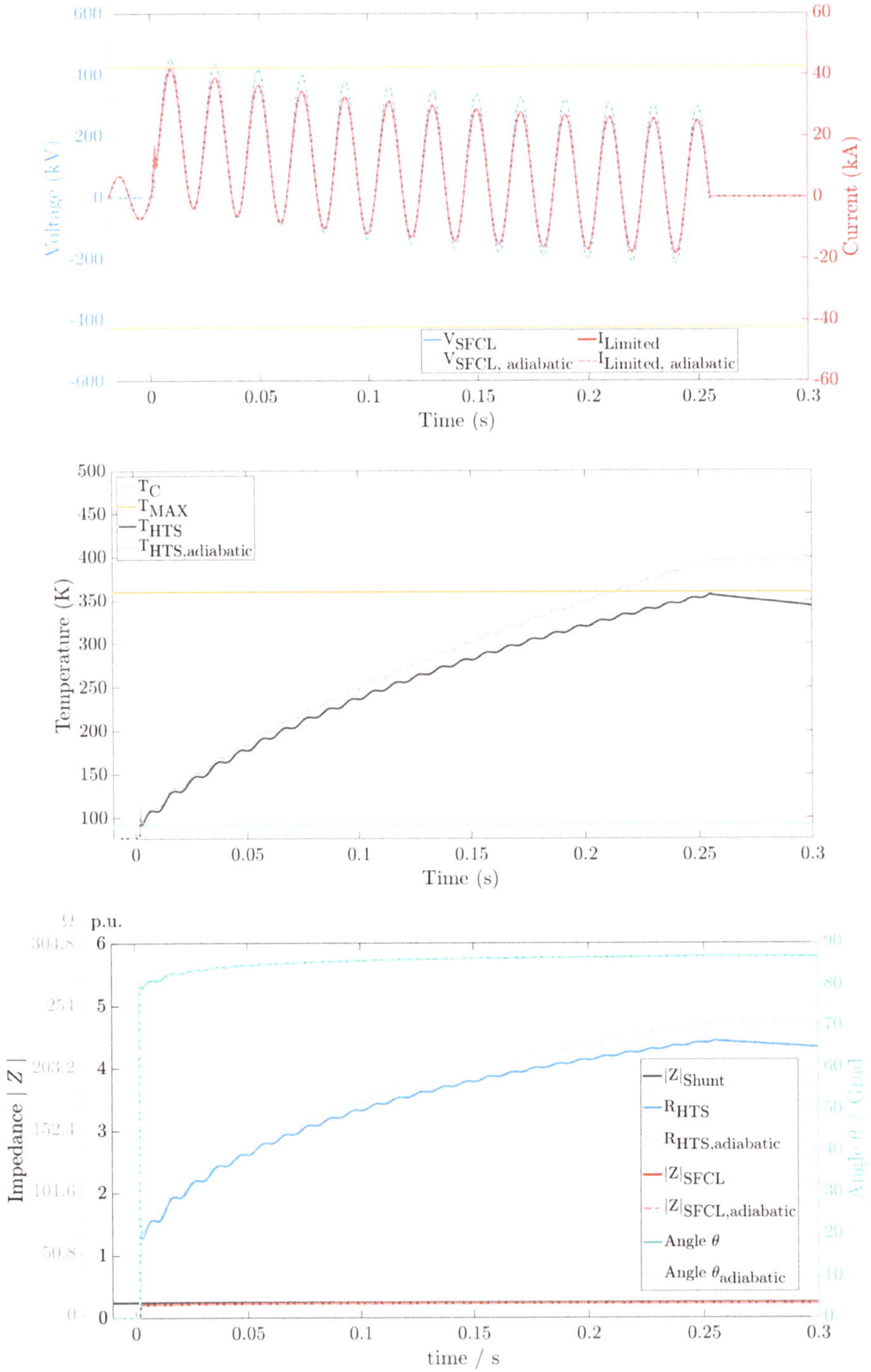

Figure 4.2.2: Current-Voltage curve (above), temperature curve of the HTS material (middle) and impedance curve of the limiter for Case B

4.3 HTS Module Design

4.3.1 Introduction

In case A, 78.4 km and in case B 69.7 km of the 12 mm wide HTS tape must be arranged per phase, with 12 conductors in parallel. There are several geometric shapes in which to arrange the superconductor as compact as possible and at the same time with good cooling. Because of the rather simple manufacturing, the negligible inductance, the compensation of magnetic fields, the low AC losses and the simple design of parallel conductors, the arrangement of the superconductor in the form of a bifilar or multifilar coil is in general the favoured design for resistive current limiters [Kraemer2007, Elschner2012]. Figure 4.3.1 shows a schematic sketch and a designed limiter module for a resistive medium-voltage current limiter.

Figure 4.3.1: Schematic diagram of a bifilar coil (left) and a designed current limiter module with high-temperature superconductors [Kraemer2021] (middle). On the right a multifilar arrangement with three bifilar coils in parallel [Noe2012]

This design can be further refined by deploying a number of tapes in parallel, with alternate current direction in adjacent tapes (multifilar design, Fig. 4.3.1, left). The number p_l is chosen according to the nominal current. This design has the advantage, that the single piece length of a tape is shorter than in a purely bifilar design. This reduces the voltage in the current limitation mode. In a series connection of modules, the outputs of a multifilar coil are then contacted to the inputs of the next coil. All tapes of a coil join on a central contact, thus allowing a redistribution of currents in the case of inhomogeneous critical current.

In a purely bifilar design the nominal current is obtained by a connection of several coils in parallel. The total need of superconducting material in both concepts remains the same.

An additional option to improve the design is the use of pairs of tapes tightly pressed together ('twins', [Elschner2012]) and connected in parallel on the same input or output contacts. This corresponds to a doubling of the critical current and thus renders the design more compact. However, the use of twins increases the magnetic field between the tapes and thus also increases the AC losses per tape. Therefore, the use of more than two tapes (triplets etc.) is not favourable.

An electrical insulation and spacer material, shown as dashed line, are co-wound between the forward conductor (blue in Fig. 4.3.1) and the return conductor (red) which insulates the voltages between the conductors. This insulation material is corrugated, allowing good flow and cooling with liquid nitrogen.

The main variable quantities in the design are the inner d_i and the outer diameter d_a of the coil, the distance between the forward and return conductor d_{tt} (distance turn to turn) and the number of parallel conductors p_l. The total length of the superconductor per bifilar coil l_{tot} is calculated using the following set of equations of an Archimedean spiral.

$$n = \frac{d_o}{2 \cdot 2\, d_{tt}} \qquad (4.3.1)$$

$$m = \frac{d_i}{2 \cdot 2\, d_{tt}} \qquad (4.3.2)$$

$$a = \frac{2\, d_{tt}}{2\, \pi} \qquad (4.3.3)$$

$$\Phi_n = 2\, \pi\, n \qquad (4.3.4)$$

$$\Phi_m = 2\, \pi\, m \qquad (4.3.5)$$

$$l_n = \frac{a}{2}\left(\Phi_n \sqrt{1 + (\Phi_n)^2} + ln\left(\Phi_n + \sqrt{1 + (\Phi_n)^2}\right)\right) \qquad (4.3.6)$$

$$l_m = \frac{a}{2}\left(\Phi_m \sqrt{1 + (\Phi_m)^2} + ln\left(\Phi_m + \sqrt{1 + (\Phi_m)^2}\right)\right) \qquad (4.3.7)$$

$$l_{tot} = 2\,(l_n - l_m) \qquad (4.3.8)$$

With *n*, *m* and *a* as parameters used for the calculation. When designing the bifilar coil of an HTS current limiting module, several boundary conditions and, in some cases, conflicting aspects must be taken into account:

- The inner diameter d_i should not be less than about 20 cm for practical reasons of contacting.
- The outer diameter d_a should be as small as possible so that the cryostat dimensions do not become too large.
- The distance between the forward and return conductors d_{tt} should be as small as possible to reduce the AC losses of the superconductor. For details see section 4.6.1.
- The distance between the forward and return conductors d_{tt} should not be less than a minimum distance for good cooling.
- In order to better isolate the voltage between the forward and return conductors, a larger distance d_{tt} between the forward and return conductors is generally preferred.
- As the length of the forward and return conductors increases, the maximum voltage between the forward and return conductors increases. This is especially true for the fast transient overvoltages. Therefore, a short length of the forward and return conductors is generally preferred.
- In order to reduce the number of bifilar coils and thus the length of the cryostat, longer forward and return conductor lengths are preferred.
- The total number of bifilar or multifilar coils in series should be high enough to keep the voltage at the coil terminal within a reasonable range.

Table 4.3.1 summarizes the total single tape length of one bifilar coil for different main parameters. More tapes in parallel reduce the single tape length and consequently the maximum voltage at the terminals of a module. The smaller this value, the easier it is to get the material from the different manufacturers because they have a higher throughput for smaller single tape length. Nevertheless, values between 50 and 100 m should be available from all manufacturers. The higher this value, the higher is the voltage at the terminals of the bifilar coil. This is discussed later in this chapter. At present, only values larger than approximately 100 m are excluded because of availability reasons.

Table 4.3.1: Tape length of one bifilar coil for inner diameter d_i of 200 mm and different values of parallel tapes p_l, outer coil diameter d_o and distance turn to turn d_{tt}

$p_l = 1$	$d_{tt} = 2$ mm	$d_{tt} = 3$ mm	$d_{tt} = 4$ mm	$d_{tt} = 6$ mm
$d_o = 800$ mm	235.6 m	157.1 m	117.8 m	78.6 m
$d_o = 1000$ mm	376.9 m	251.3 m	188.5 m	125.7 m
$d_o = 1200$ mm	549.7 m	366.5 m	274.9 m	183.3 m
$p_l = 2$				
$d_o = 800$ mm	117.8 m	78.5 m	58.9 m	39.3 m
$d_o = 1000$ mm	188.5 m	125.6 m	94.3 m	62.4 m
$d_o = 1200$ mm	247.9 m	183.2 m	137.5 m	91.7 m
$p_l = 3$				
$d_o = 800$ mm	78.6 m	52.3 m	39.2 m	26.1 m
$d_o = 1000$ mm	125.7 m	83.8 m	62.8 m	41.9 m
$d_o = 1200$ mm	183.3 m	122.2 m	91.6 m	61.1 m
$p_l = 4$				
$d_o = 800$ mm	58.9 m	39.2 m	29.5 m	19.5 m
$d_o = 1000$ mm	94.3 m	62.8 m	47.1 m	31.4 m
$d_o = 1200$ mm	137.5 m	91.6 m	68.7 m	45.9 m
$p_l = 6$				
$d_o = 800$ mm	39.3 m	26.1 m	19.5 m	13.1 m
$d_o = 1000$ mm	62.4 m	41.9 m	31.4 m	20.9 m
$d_o = 1100$ mm	76.6 m	51.0 m	38.3 m	25.5 m
$d_o = 1200$ mm	91.7 m	61.1 m	45.9 m	30.6 m

4.3.2 Design Options and Preferences

The main objective of this section is to find a suitable set of parameters for the HTS current limiting element that enables a compact design, a low voltage and still acceptable AC losses. For this reason, the main parameters vary within reasonable ranges. The selection was made based on the following assumptions:

- The total limiting element length should be less than 7 m in order to limit the total length of the cryostat.
- The maximum voltage for one HTS module at current limitation should be less than 20 kV for a homogenous voltage distribution.
- The distance d_{tt} between the turns was set to 2 mm in order to get reasonable AC losses.
- The outer diameter of one HTS module varies between 800 mm and 1200 mm in order to achieve reasonable dimensions.

- Twelve HTS tapes were connected in parallel to carry the current. Single and twin tape arrangements were investigated.
- A critical current of 700 A is assumed to reduce the total amount of HTS material.

Table 4.3.2 and Table 4.3.3 show the results for the number of HTS modules, the voltage during limitation and the total length of the limiting element for case A and case B. For both cases, only twin tape arrangements with an outer diameter of 1100 mm for case A and 1000 mm for case B fulfil the requirements stated above. The parameters highlighted in green were selected for the further design process.

Table 4.3.2: HTS module and limiting element parameters for case A, d_{tt}=2, 78.4 km, 6.5 km in series, 700 A

d_i = 200 mm	Tape length in a single bifilar coil	Number of modules in series connection	Number of modules in total	Module arrangement	Total limiting element length	Voltage
p_l = 3, 12 single tapes in parallel						
d_o = 800 mm	78.6 m	83	83 · 4	(4·25+50)mm·83	12.45 m	18.9 kV
d_o = 1000 mm	125.7 m	52	52 · 4	(4·25+50)mm·52	7.8 m	30.1 kV
d_o = 1200 mm	183.3 m	36	36 · 4	(4·25+50)mm·36	5.4 m	43.6 kV
p_l = 3, 6 twin tapes in parallel						
d_o = 800 mm	78.6 m	83	83 · 2	(2·25+50)mm·83	8.3 m	18.9 kV
d_o = 1000 mm	125.7 m	52	52 · 2	(2·25+50)mm·52	5.2 m	30.1 kV
d_o = 1200 mm	183.3 m	36	36 · 2	(2·25+50)mm·36	3.6 m	43.6 kV
p_l = 4, 12 single tapes in parallel						
d_o = 800 mm	58.9 m	111	111 · 3	(3·25+50)mm·111	13.875 m	14.1 kV
d_o = 1000 mm	94.3 m	69	69· 3	(3·25+50)mm·69	8.625 m	22.8 kV
d_o = 1200 mm	137.5 m	48	48 · 3	(3·25+50)mm·48	6.0 m	32.7 kV
p_l = 6, 12 single tapes in parallel						
d_o = 800 mm	39.3 m	166	166 · 2	(2·25+50)mm·166	16.6 m	9.5 kV
d_o = 1000 mm	62.4 m	105	105 · 2	(2·25+50)mm·105	10.5 m	15.0 kV
d_o = 1200 mm	91.7 m	71	71 · 2	(2·25+50)mm·71	7.1 m	22.1 kV
p_l = 6, 6 twin tapes in parallel						
d_o = 800 mm	39.3 m	166	166	(25+50)mm·166	12.45 m	9.5 kV
d_o = 1000 mm	62.4 m	105	105	(25+50)mm·105	7.875 m	15.0 kV
d_o = 1100 mm	76.6 m	85	85	(25+50)mm·85	6.375 m	18.5 kV
d_o = 1200 mm	91.7 m	71	71	(25+50)mm·71	5.325 m	22.1 kV

Table 4.3.3: HTS module and limiting element parameters for case B, d_{tt}=2 mm, 69.7 km, 5.8 km in series, 700 A

d_i = 200 mm	Tape length in a single bifilar coil	Number of modules in series connection	Number of modules in total	Module arrangement	Total limiting element length	Voltage
p_l = 3, 12 single tapes in parallel						
d_o = 800 mm	78.6 m	74	74 · 4	(4·25+50)mm·74	11.1 m	21.2 kV
d_o = 1000 mm	125.7 m	47	47 · 4	(4·25+50)mm·47	7.05 m	33.4 kV
d_o = 1200 mm	183.3 m	32	32 · 4	(4·25+50)mm·32	4.8 m	49.0 kV
p_l = 3, 6 twin tapes in parallel						
d_o = 800 mm	78.6 m	74	74 · 2	(2·25+50)mm·74	7.4 m	21.2 kV
d_o = 1000 mm	125.7 m	47	47 · 2	(2·25+50)mm·47	4.7 m	33.4 kV
d_o = 1200 mm	183.3 m	32	32 · 2	(2·25+50)mm·32	3.2 m	49.0 kV
p_l = 4, 12 single tapes in parallel						
d_o = 800 mm	58.9 m	99	99 · 3	(3·25+50)mm·99	12.37 m	15.9 kV
d_o = 1000 mm	94.3 m	62	62 · 3	(3·25+50)mm·62	7.75 m	25.3 kV
d_o = 1200 mm	137.5 m	43	43 · 3	(3·25+50)mm·43	5.3 m	36.5 kV
p_l = 6, 12 single tapes in parallel						
d_o = 800 mm	39.3 m	148	148 · 2	(2·25+50)mm·148	14.8 m	10.6 kV
d_o = 1000 mm	62.4 m	93	93 · 2	(2·25+50)mm·93	9.3 m	16.9 kV
d_o = 1200 mm	91.7 m	64	64 · 2	(2·25+50)mm·64	6.4 m	24.5 kV
p_l = 6, 6 twin tapes in parallel						
d_o = 800 mm	39.3 m	148	148	(25+50)mm·148	11.1 m	10.6 kV
d_o = 1000 mm	62.4 m	93	93	(25+50)mm·93	6.97 m	16.9 kV
d_o = 1200 mm	91.7 m	64	64	(25+50)mm·64	4.8 m	24.5 kV

4.4 High Voltage Design

The general setup and the main components of one phase of the limiter is shown in Fig. 4.4.1. All areas with high field stress are numbered and an estimation of the maximum field stress is determined in the following positions:

1) The maximum E-field between the outer shielding ring of the limiting element and the inner wall of the cryostat at ground potential is in liquid nitrogen at subcooled conditions with 77 K and 5 bar.
2) The maximum E-field at the end of the limiting element and the cryostat head at ground potential. The insulation material is liquid nitrogen at subcooled conditions with 77 K and 5 bar.
3) The bushing at cryogenic conditions needs a grading to control the maximum E-field. We assume that the grading area is within liquid nitrogen at 77 K and 5 bar.
4) The maximum E-field within a bifilar limiting element needs a careful design because during limitation the occurrence of gaseous nitrogen as a dielectric needs to be considered.
5) During fast voltage transients the series connection of several HTS modules can lead to non-linear voltage distribution along the modules.
6) The support structure for the mechanical support of the current limiting element is located in liquid nitrogen at 77 K and 5 bar.

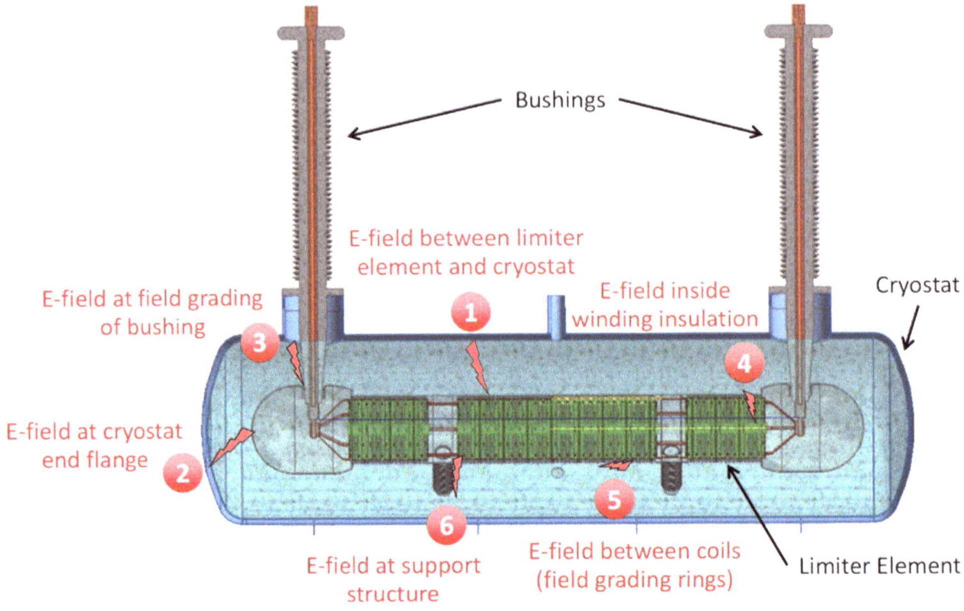

Figure 4.4.1: General scheme of the cryostat with limiter elements and areas with high voltage design aspects investigated within this study

4.4.1 Introduction to High Voltage Design

The main electrical insulation material for superconducting fault current limiters is liquid nitrogen, which has very good electrical insulation properties [Gerhold2002]. Figure 4.4.2 shows experimentally determined breakdown voltages [CIGRE2021]. While many investigations have been carried out for voltages below 100 kV, only few investigations for voltages above 100 kV are reported. The highest impulse voltage of about 500 kV was obtained at an electrode distance of less than 20 mm.

Our own investigations [Fink2014] show that the breakdown voltages of liquid nitrogen scatter significantly more with increasing electrode distance and, as can also be seen in Fig. 4.4.2, increase only slightly for non-uniform fields. Furthermore, it should be noted that the breakdown voltage depends on a number of other parameters, such as volume, polarity, surface roughness, voltage shape (pulse, AC), purity of the insulating liquid, transition points (triple point), aging and pressure.

In resistive current limiters, bubbles occur during the limiting process due to the heating of the superconductor, which further reduce the dielectric strength considerably [Hara1989]. Then, the process of a gas breakdown dominates the breakdown mechanism. The Paschen curve for gaseous nitrogen is shown in Fig. 4.4.3 [Dakin1970].

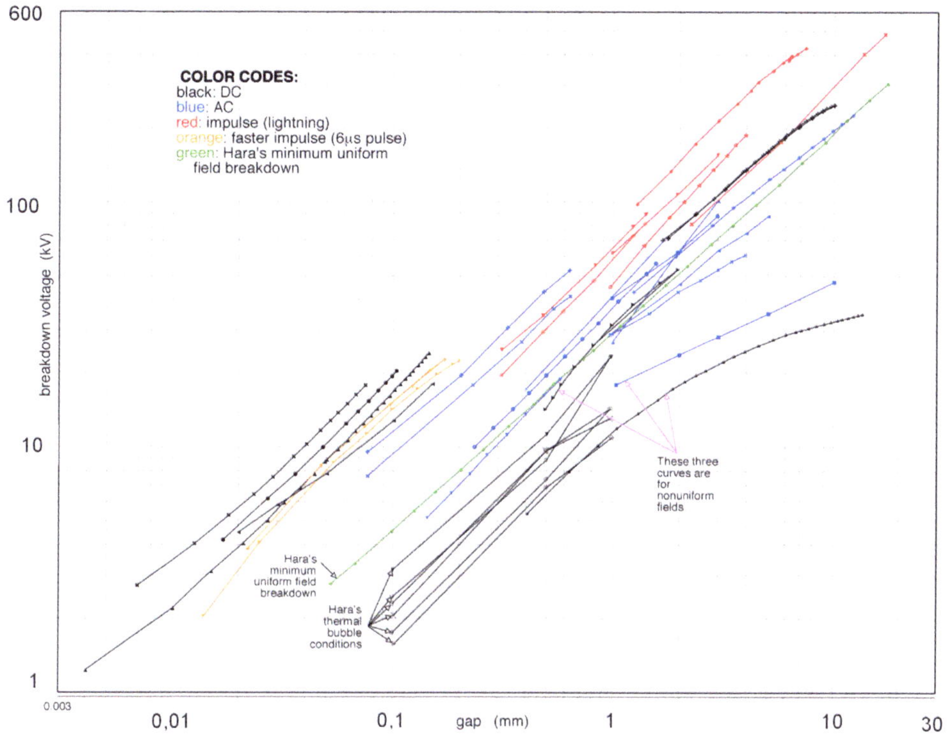

Figure 4.4.2: Breakdown voltages of liquid nitrogen according to [CIGRE2016]

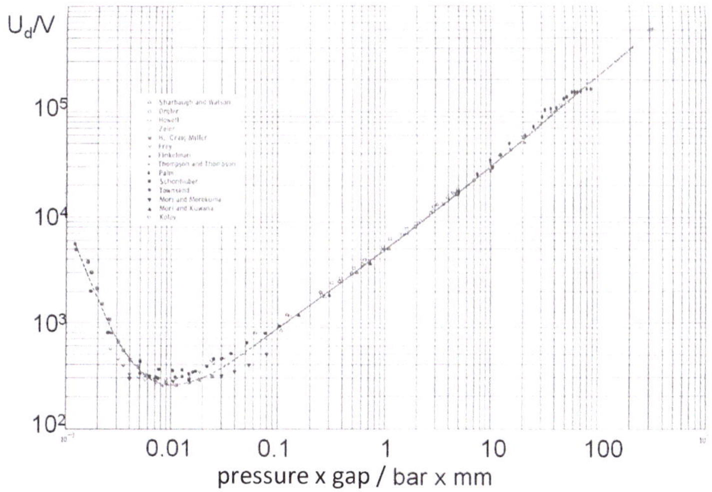

Figure 4.4.3: Paschen curve for nitrogen at room temperature [Dakin 1970]

The breakdown voltage during the formation of bubbles in liquid nitrogen has been investigated only at low electrode distances and stresses. Since there is a significant reduction compared with the pure liquid, the conservative assumption is made in this investigation that the gaseous nitrogen is present as an insulating medium. This should provide a reliable cryogenic high voltage design and the following calculations are based on this assumption.

4.4.2 Maximum Electric Field Calculation

4.4.2.1 Derivation of Maximum Electric Field and Liquid Nitrogen Pressure

For a homogeneous electric field and normal conditions (1 bar, 25 °C), a breakdown field strength of roughly 24.5 kV/cm results in nitrogen. With a safety factor of two and a dielectric field constant of liquid nitrogen of ε_r = 1.43, the maximum field strengths and insulation distances listed in Table 4.4.1 result. As can also be seen in the Paschen curve in Fig. 4.4.3, the insulating distance can be reduced considerably by increasing the pressure in the cryostat.

Table 4.4.1: Maximum electric field and insulation distance in gaseous nitrogen (Assumptions: Safety factor 2, dielectric field constant 1.43)

Pressure in cryostat	Breakdown electric field	Max. electric field	Distance for homogenous field at 1570 kV
1 bar	1.713 kV/mm	0.856 kV/mm	1834.1 mm
3 bar	5.139 kV/mm	2.564 kV/mm	612.3 mm
5 bar	8.566 kV/mm	4.283 kV/mm	366.5 mm

For a nominal voltage of 380 kV it is clear that the pressure in the cryostat must be chosen to be higher than 1 bar otherwise, the cryostat diameter would be unreasonably high. For this study, a liquid nitrogen pressure of 5 bar is chosen to reduce the cryostat diameter to a reasonable value. The 220 kV limiter realized by SuperOx [Moyzikh2021] in Moscow operates the liquid nitrogen at 5 bar and assumes a maximum electric field of 4.8 kV/mm at the corona rings and 9.7 kV/mm in the area of the bushing (see. Fig. 9.2.4).

The maximum voltage of 1570 kV was selected according to the test voltage for chopped standard lightning impulse (LIC) for a maximum voltage of equipment U_m = 420 kV (Table 2.2.2) in combination with the maximum field values in Table 4.4.2. In principle, other waveform voltages with other maximum field limits can also be used. The selection of the chopped standard lightning impulse voltage as maximum voltage was made for several reasons. Most importantly, the material for the investigated design is liquid nitrogen. Tests with full wave standard lightning impulse (LI) at KIT with the facility Fatelini 2 with liquid nitrogen showed large scattering of time to breakdown, so that in some cases time to breakdown values occurred as specified for chopped standard lightning impulse. Hence, the higher voltage of a chopped standard lightning impulse was selected compared to a standard lightning impulse. In addition, breakdown values for liquid nitrogen with Fatelini 2 and different waveforms make it useful to select a standard lightning impulse instead of an AC withstand voltage even by considering the ratio of test voltages between a standard lightning impulse and an AC withstand voltage. This is even valid if the AC peak value is replaced by the higher peak value of the standard switching impulse (SI).

In a first step, a field strength limit of 4.0 kV/mm was selected for one phase liquid nitrogen considering the different information mentioned above. This is in agreement with the limit for bubble free liquid nitrogen selected for the field calculation for a 154 kV SFCL published in [Kim2015]. In the final step a reduction of the field strength for pure liquid nitrogen to 3.8 kV/mm was done to consider a possible effect of the larger volume of the 380 kV SFCL. Further reduction by influence of gas bubbles was not anticipated between the high voltage shielding rings around the modules and the grounded cryostat wall because of the short time to switch off and – even more important – because of the subcooled conditions of liquid nitrogen.

Table 4.4.2: Maximum electric field in different dielectric materials

Dielectric	T / K	E_{max} / kV/mm	Comments
LN_2	77	3.8	Consider large volume
$LN_2 + N_2$	up to 300 for gas	2.5	
$LN_2 + N_2$	360	2.08	360 / 300 = 1.2 (correction formula for air only valid until 1.1)
N_2	300	3.8	Not better than LN_2 No large volume (no leader)
LN_2 / FRP	77	2.7	selected from data in [Koo2017]
FRP	77	5.0	selected from data in [Tuncer2009]

4.4.2.2 Estimation of Cryostat Inner Diameter

The current limiter module and the cryostat can be considered as shown in Fig. 4.4.4 simplified as a coaxial arrangement, and with the help of the utilization factors η of Schwaiger [Schwaiger1925] the isolation distance for inhomogeneous field arrangements can then be estimated. As Fig. 4.4.4 and the following equations show, that with a clear distance of around 75 cm and a maximum electric field of 3 kV/mm it is possible to keep the diameter of the cryostat in the range of 3 m.

$$p = \frac{s + r}{r} = \frac{750 + 750}{750} = 2 \qquad (4.4.1)$$

$$E_{max} = \frac{U}{\eta\ s} = \frac{1570\ \text{kV}}{0.693 \cdot 750\ \text{mm}} = 3.002\ \frac{\text{kV}}{\text{mm}} \qquad (4.4.2)$$

η is the Schwaiger utilization factor for this geometry.

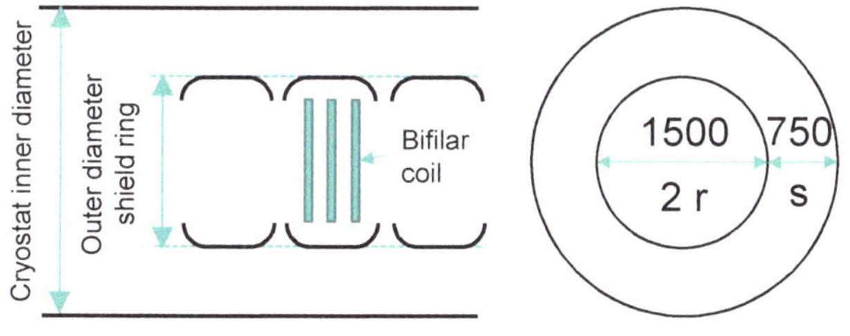

Figure 4.4.4: Basic arrangement of modules inside cryostat

To summarize, with a liquid nitrogen pressure of 5 bar and a maximum electric field of 3 kV/mm a reliable high voltage design with reasonable dimensions of the cryostat should be possible. For safety reasons and the unknown experimental breakdown voltages at such high voltages the maximum electric field is chosen lower than in the 220 kV limiter realized by SuperOx.

4.4.3 High Voltage Design outside Modules based on FEM Calculation

In this chapter, a tentative design is outlined for the space outside the modules. Comsol Multiphysics 5.5 was used for field calculation. A sketch of one "geometry" (Comsol wording) is shown in Fig. 4.4.5.

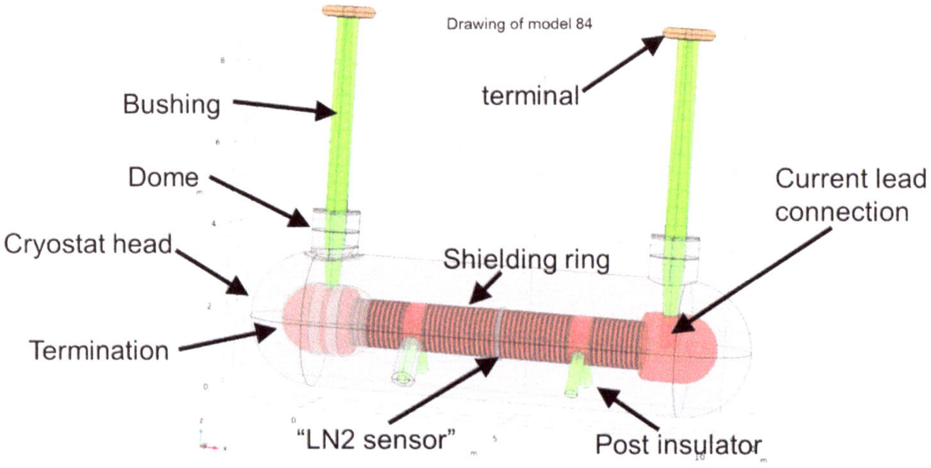

Figure 4.4.5: Drawing of Comsol model 84 with cryostat, bushings, terminations and shielding rings around modules. (Spaces with high mesh density for obtaining high accuracy of the field strength are called "sensors")

The models are based on different assumptions, where voltage values are kept fixed. Shapes and dimensions were adjusted in order to satisfy maximum field strength limits by keeping the size as low as possible. During this calculation process, the cryostat diameter size was a dominant parameter and it increased in steps derived from standardized manufacturing sizes for pressure vessel heads. It was assumed that the vessel diameter for heads and straight cylinder parts was the same. New adjustment and calculations were then performed until the maximum field values were satisfied. Calculation effort reduction by using symmetry properties is possible especially for the assumption that all high voltage parts (excluding the floating layers of the bushings) are on the same high voltage level. For several reasons (e.g. to allow later different voltages on different shielding rings) such a simplification was not done. A full size model for all regions would be too large to handle with the available hardware restriction of 64 GB RAM. This is solved by making smaller models, each representing a special region of the full size model.

4.4.3.1 Cryostat Head and Straight Cylinder

Based on initial estimations according to previous chapters, a minimum value of 3 m for the inner diameter was used as starting value for the cylindrical part. A thickness of 10 mm for the cryostat wall was derived from a rough estimation based on pressure vessel regulations. Selection of dimensions according to standard sizes for pressure vessel heads and an outer wall value of 3200 mm led to an inner diameter of 3180 mm.

Taking field increase on the terminations into consideration it was necessary to further increase the cryostat diameter as shown in Table 4.4.3. The minimum possible diameter size of 3380 mm caused a maximum field value of 3.72 kV/mm for the best case when the inner sphere has half of the diameter of the cryostat inner wall.

Based on this mathematical background it was concluded that the minimum design size can be obtained for a profile consisting of a hemi-spherical cryostat head and a spherical termination with a transition to a cylinder-shaped arrangement of HTS modules between the termination spheres.

Table 4.4.3: Comparison between cylindrical and spherical field. The abbreviations d_2, r_1 and r_2 are valid for the outer diameter and inner and outer radius, respectively. U = 1570 kV

	Coaxial cylinders	Spheres
	$E(\mathbf{r}) = \frac{U}{r \cdot \ln {}^{r_2}/_{r_1}}$	$E(\mathbf{r}) = \frac{1}{r^2} \cdot \frac{U}{\frac{1}{r_1} - \frac{1}{r_2}}$
$r_2 = \infty$	$E(r_1) = 0$	$E(r_1) = U/r_1$ $E(r_1$=0.5 m) = 3.14 kV/mm $E(r_1$=0.7 m) = 2.24 kV/mm
Condition of minimum field strength at constant r_2	$r_2/r_1 = e$	$r_2/r_1 = 2$
E_{opt} for $d_2 = 2\,r_2$ = 3.38 m	2.53 kV/mm	3.72 kV/mm
E_{max} for different r_1 with d_2 = 3.38 m	$E(r_1$=0.5 m) = 2.58 kV/mm $E(r_1$=0.7 m) = 2.54 kV/mm	$E(r_1$=0.5 m) = 4.46 kV/mm $E(r_1$=0.7 m) = 3.83 kV/mm

The bushings can be placed on different locations near the terminations. A short length is obtained by placing the bushing above the termination sphere with the disadvantage of an increased dome height. FEM calculations confirm that maximum fields of this version do not exceed the allowed values. Another version with the bushings positioned sideward from the termination sphere (towards the inner side of the SFCL) allows minimum dome height. The disadvantage is an increase in cryostat length and electrical field values obtained above the limits in a first version. Hence, this concept was not further investigated. Dome height reduction to 1.1 m was agreed by 3D calculation for the model with bushings above each termination.

Field strength transgression may occur in real world conditions by marginal deviation from the theoretical sphere-to-sphere design, e.g. by manufacturing tolerances, surface changes (welding, screw head) thermal contraction and other reasons. An example for a slight surface change was calculated and resulted in an exceeding of the allowed field value. A possible solution proposes a model where the termination sphere splits in two hemispheres and a cylindrical part added between the hemispheres. A model was calculated and confirmed for dome height reduction to 1.1 m too. The sketch in Fig. 4.4.5 shows the terminations including the cut in the cylindrical part, which allows the lowering of the bushing. On the surface of the sphere there are now regions with lower field which allow changes from the perfect shape, e.g. for assembly purposes.

A mechanical challenge was the weldseam collision for dome and cryostat head appearing for the modelled setup with hemispherical heads as shown in Fig. 4.4.6.

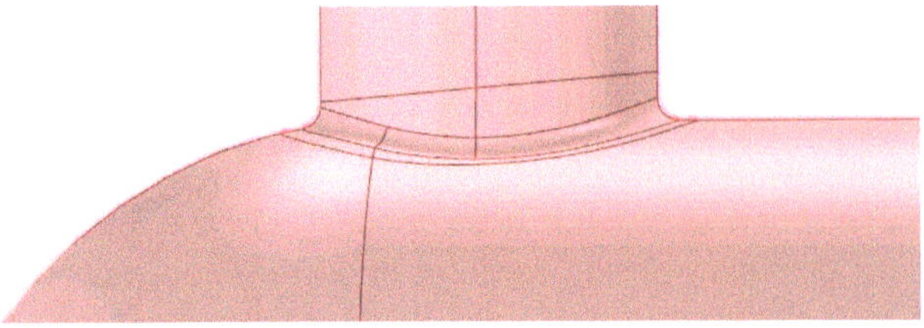

Figure 4.4.6: Collision of weldseams. One weldseam is between the hemispherical head and cylindrical part of the cryostat. The other weldseam is between cryostat and dome.

The collision can be avoided by shifting the hemisphere away from the dome in horizontal direction, elongate the cylindrical part for the shifted length (to maintain the contact between cryostat and head) and keeping the dome on the same position. The disadvantage of this action is a longer total length of the cryostat. Replacing the hemisphere head by an ellipsoidal or torispherical head may compensate the total length elongation (Fig. 4.4.7).

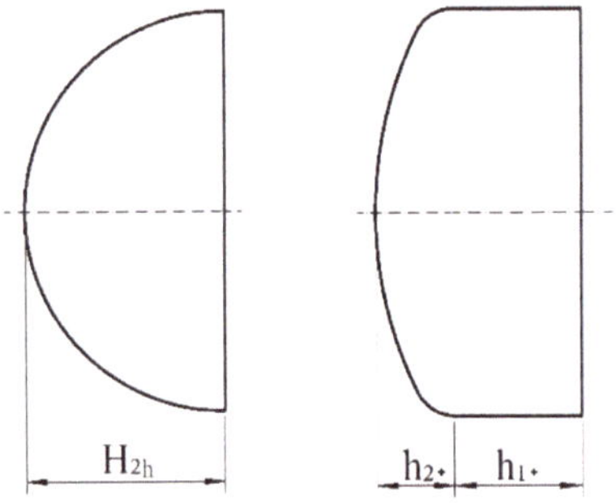

Figure 4.4.7: Replacing of a hemispherical head of the length H_{2h} by a combination of a torispherical head of the length h_{2+} with a cylinder of the length h_{1+}. The addition of the cylinder avoids a collision between dome and head seamwelds.

The additional horizontal length of the cylindrical part must not only ensure that the setup is seamweld collision free. The additional cylindrical length must also ensure that the field strength limits will not be exceeded because the other heads have smaller radii than the hemispherical head. Therefore, for a first field calculation, the necessary respective horizontal length elongation for setups with ellipsoidal and torispherical head were calculated which resulted in the same total length as the setup with hemispheric heads (Table 4.4.4). An agreement of the cylinder and torispherical head combination length with the length of the hemisphere head shows Fig. 4.4.7.

Table 4.4.4: Elongation of cylindrical part of cryostat

Cryostat head type	Hemisphere	Ellipsoidal	Torispherical
Termination head type	Hemisphere	Hemisphere	Hemisphere
Data	inner diameter of cryostat inner wall d = 3.380 m, outer diameter of cryostat inner wall d_a = 3.400 m, cryostat inner wall thickness s = 10 mm		
Standard		DIN 28011	DIN 28013
	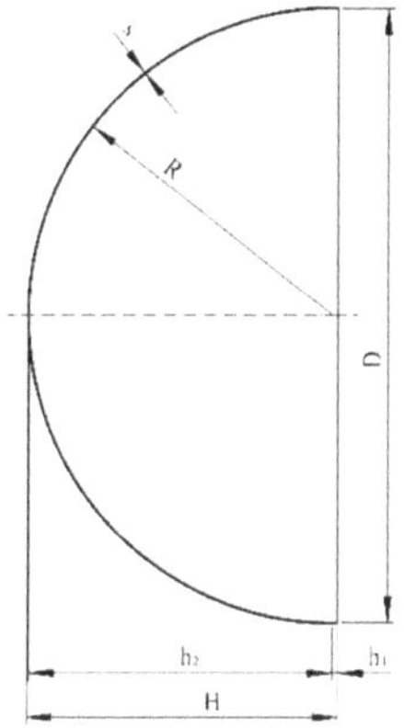	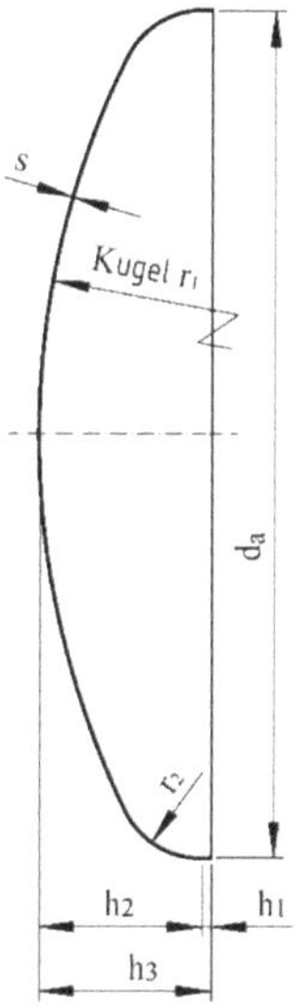	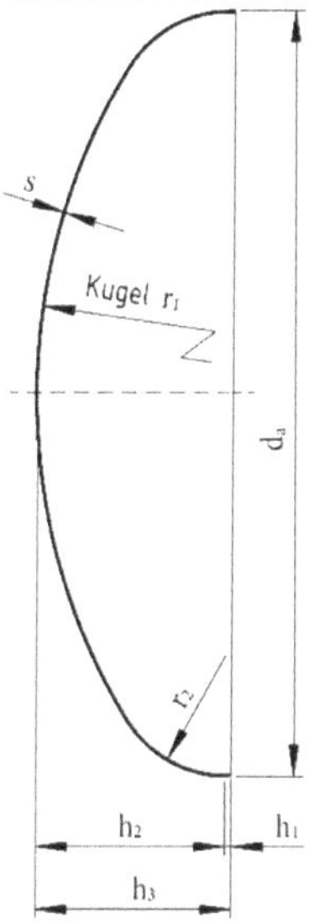
Termination head radius	$d/4$	$d/4$	$d/4$
Cryostat head radius	$d/2$ = 1.690 m		
h_2 according to standard		0.255 d_a – 0.635 s = 0.86065 m	0.1935 d_a – 0.455 s = 0.65335 m
elongation of cryostat cylindrical part at one side to keep total cryostat length: $d_l = d/2 - h_2$	0	1.690 m - 0.86065 m = 0.82935 m	1.690 m - 0.65335 m = 1.03665 m
Distance of cryostat head and termination head at the rotation axis $r = 0$	$d/2 - d/4 = d/4$ = 0.845 m	0.845 m	0.845 m

FEM-calculations with simplified 2D models were done with the values d_l for the elongation of the cryostat cylindrical part to keep the total cryostat length. Since it turned out that the field strength limit of 3.8 kV/mm was not reached for both alternative head shapes, other lengths of the cryostat were also investigated. For the torispherical head setup the maximum fields are shown for different horizontal elongations in Table 4.4.5. The electrical field strength for the additional cylindrical length of 1.04 m shows the last row of Table 4.4.5. The 1.04 m is according to Table 4.4.4 the length for the torispherical head setup, which results in the same length as the hemisphere setup. The maximum field for the 1.04 m elongation is 3.56 kV/mm and is lower than the limit of 3.8 kV/mm. According to Table 4.4.5, the shortest elongation which satisfies the 3.8 kV/mm field limit can be found for 0.92 m which is 12 cm less than the length of 1.04 m.

Table 4.4.5: Additional horizontal length "cryostat_dl" for torispherical head setup to avoid weldseam collision and resulting maximum field strength "Emax termination"

cryostat_dl / m	Emax termination / kV/mm
0.80000	4.1924
0.89665	3.8503
0.91665	3.7992
0.93665	3.7529
0.98665	3.6540
1.0267	3.5794
1.0367	3.5609

It was calculated that for an ellipsoidal head a reduction of 9 cm and for the torispherical head a reduction for 12 cm would be possible per side. The maximum possible total length reduction would be twice of these values for each.

4.4.3.2 Shielding Rings for Modules

Shielding rings around the modules ensure a smooth high voltage surface to the grounded cryostat wall. In case the modules were connected in parallel, a common ring was used for both modules. Such a version was investigated for a triple module as initial design. Due to a design change, no HTS modules were switched in parallel and the design included one shielding ring for each module.

According to the HTS module design, a length in horizontal direction (of the "lying" cryostat tube) of 75 mm was available for each shielding ring in combination with its gap. A minimum straight length of 25 mm was assumed in order to ensure a straight overlapping of the module. A torus with its cross section of a circle may be a preferable shape from the manufacturing point of view but it would lead to too high field values.

Figure 4.4.8 left shows one shielding ring including a straight part instead of a perfect circular cross section and Fig. 4.4.8 right shows a cross section of a shielding ring. Figure 4.4.8 right shows the composition of the cross section with the central rectangle for the straight part and additional ellipses and rectangles with fillets.

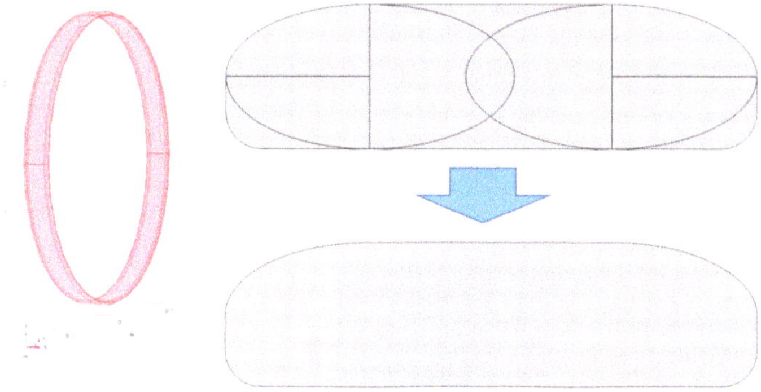

Figure 4.4.8: Left: Shielding ring for module triplet. Right top: Upper cross section of shielding ring for single module. Inner lines are visible for showing the inner structure. Right bottom: These inner lines were removed in a further step.

The reference design for the cross section of the shielding ring was:

- Straight length (i.e. length of central rectangle in Fig. 4.4.8): 25 mm
- Rounding per side, each (i.e. half of the length of each ellipse in Fig. 4.4.8): 15 mm
- Gap between two shielding rings: 20 mm

Most of the investigated 3D models were valid for an electrical short SFCL, i.e. all shielding rings, terminations and busbars were on the same high voltage value of 1570 kV. These 3D shielding ring models satisfied the limit of 3.8 kV/mm. A 2D model was built by cutting an existing 3D model and using one part for the axis symmetric 2D calculation. Unfortunately, Comsol can only show the symmetry axis in vertical position. Hence, the model is shifted by 90° (see Fig. 4.4.9). The model indicated too high fields in case of a voltage of 1570 kV on a shielding ring and 1540 kV on an adjacent shielding ring. The modules were represented by one common cylinder in order to keep the model simple.

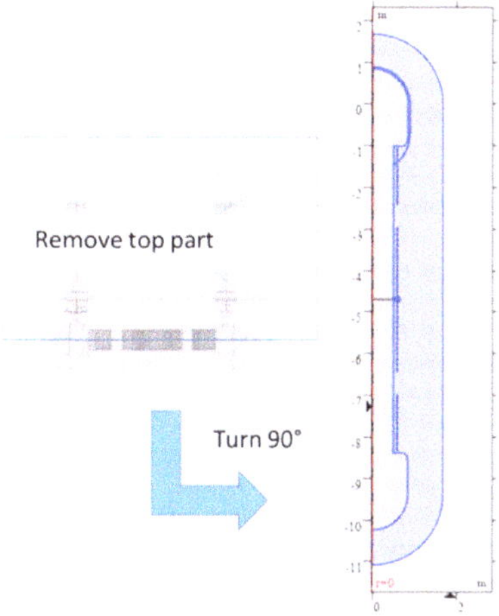

Figure 4.4.9: From 3D to 2D model. Start with 3D model. Cut upper half and remove it. Shift orientation by 90° for an axis symmetric model

For a more detailed investigation an axis symmetric 2D mini model was built (Fig. 4.4.10 left). The z-axis of this 2D model represented the x-axis of a 3D model. The r-axis of the 2D model represented the z-axis of a 3D model. The model contained 105 shielding rings. Twelve central rings were each equipped with their modules. Half of the rings and modules were on a voltage of 1570 kV. The other half were on 1540 kV, 1550 kV, 1560 kV or 1570 kV, each. In the example shown in Fig. 4.4.10 right the second voltage was 1540 kV.

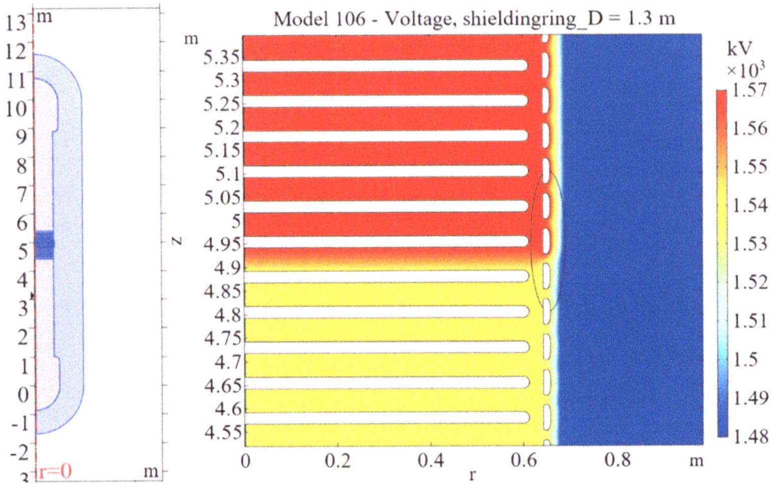

Figure 4.4.10: Left: Full view on 2D model, right: Zoom to central part of 2D model. In this example half of the rings and modules had a voltage of 1570 kV (red) and the others 1540 kV (yellow)

A geometry parameter sweep was combined with the parameter sweep of the second voltage. The geometry parameter "shieldingring_dgap" indicated the difference of the gap length compared to the reference ring design. Since the total length was assumed as fixed the straight length of the coil cross section was decreased for the same parameter "shieldingring_dgap".

A field distribution example with indication of the location for the maximum field can be seen in Fig. 4.4.11. It is significant that this location depends on the voltage difference of the rings.

A comparison of shielding ring cross section can be seen for the field plots in Fig. 4.4.12. In Fig. 4.4.13 the calculation results for the maximum electric field is shown depending on shielding ring shape and voltage of the second terminal. The voltage of the first terminal is kept at 1570 kV.

Electric field values below the limit of 3.8 kV/mm were calculated for all investigated shapes in case of a common shielding ring voltage of 1570 kV. Higher field maxima were obtained in case of lower voltages for the second half of shielding rings. The reference shape and all shapes with shorter gaps obtained values below 3.85 kV/mm for a difference of 20 kV (i.e. U_terminal2 = 1550 kV). Only for the shape with 5 mm shorter gap was a field value calculated below 3.80 kV/mm for a 20 kV difference. For a 30 kV difference (i.e. U_terminal2 = 1540 kV) all maximum field values are shown in Fig. 4.4.13 to be above 4.40 kV/mm with the reference shape obtaining the lowest value of 4.41 kV/mm.

Finally, it can be concluded, that there is no "best" shape for all cases but the optimum shape is voltage distribution dependent.

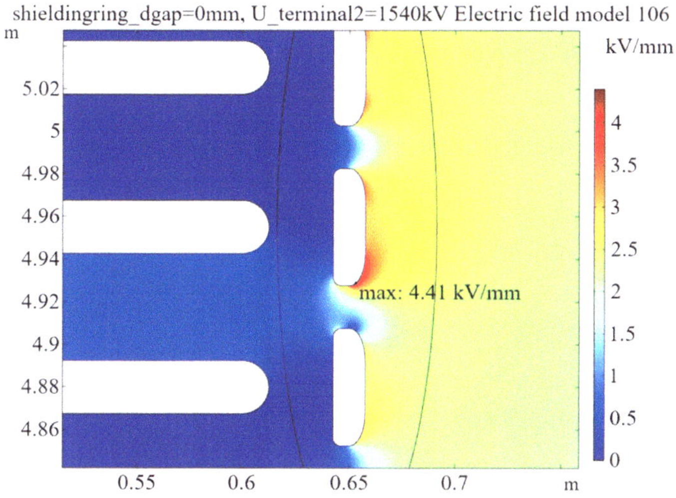

Figure 4.4.11: Electric field distribution with maximum field location

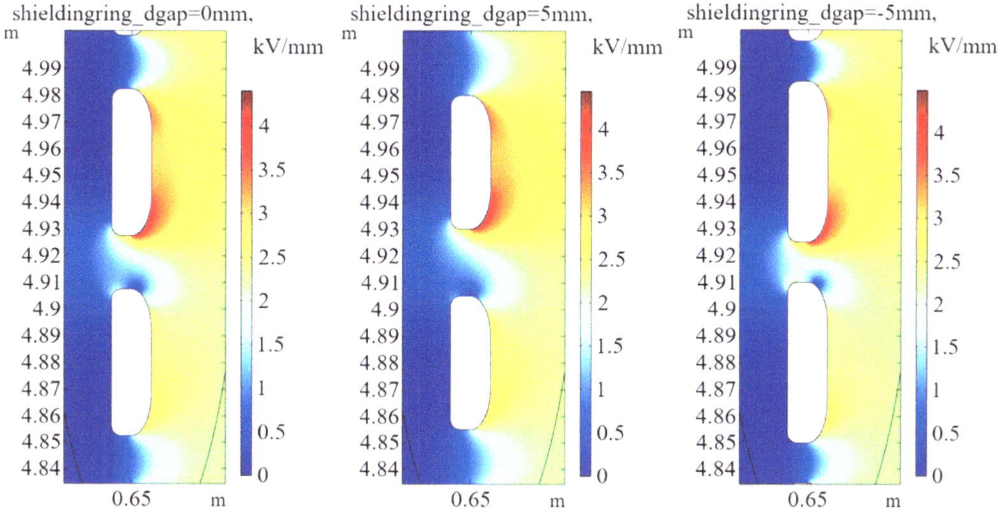

Figure 4.4.12: Electric field for the left: reference shielding ring setup, middle: for a gap difference of + 5 mm to the reference design and right: for a gap difference of – 5 mm to the reference design

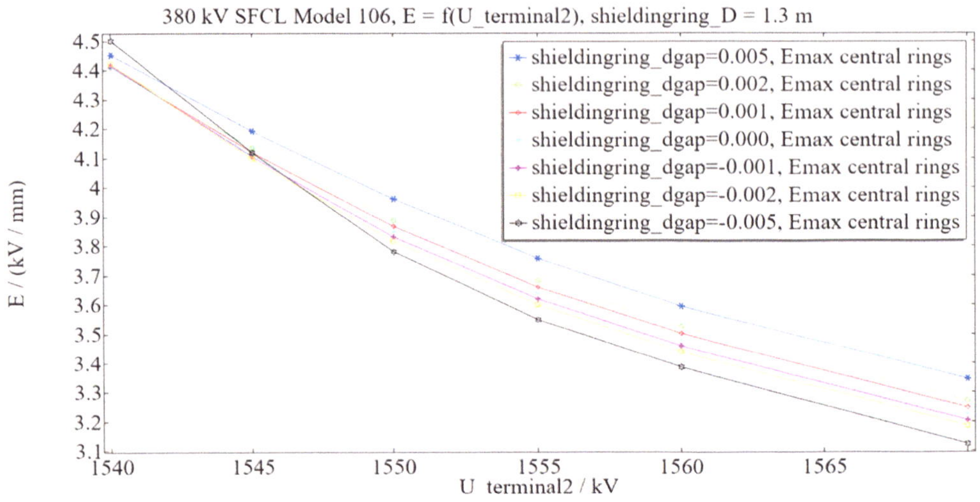

Figure 4.4.13: Maximum field vs. voltage of the second terminal for a voltage of 1570 kV for the first terminal. In this figure the parameter "shieldingring_dgap" is displayed in meter.

4.4.3.3 Shielding for Post Insulator

Post insulators were assumed to have a bipod shape as shown in Fig. 4.4.14. Two bipods were foreseen for the SFCL but for a more detailed investigation 3D mini models with only one bipod were simulated (Fig. 4.4.14). A rib-free FEM calculation delivered no field enhancement on rod surfaces. However, under experimental conditions, adding ribs often leads to higher flashover voltages. For a solid to liquid nitrogen surface such an effect is described in [Koo2017]. For our model, 8 ribs were added on the tube.

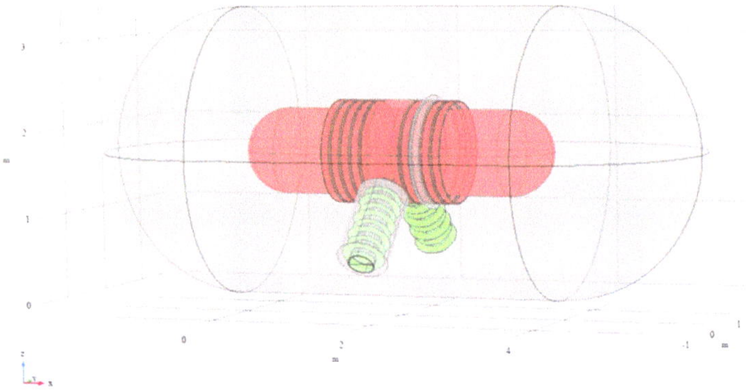

Figure 4.4.14: Model 95 bipod mini model with ribs on insulator rod

A cut plane was introduced in the model according to the red line in Fig. 4.4.15 left cutting one post insulator leg of the bipod. Figure 4.4.15 right shows the space around the shielding of this leg. The first choice of insulator material for the bipod is FRP with a permittivity of $\varepsilon_r = 3.5$. A central metal (orange color in Fig. 4.4.15 right) insert was modelled in the FRP (yellow) of the rod in order to avoid too high fields on the inner rounding of the copper shield and on the triple point. In order to save space, it is possible to integrate a module in the bipod shield.

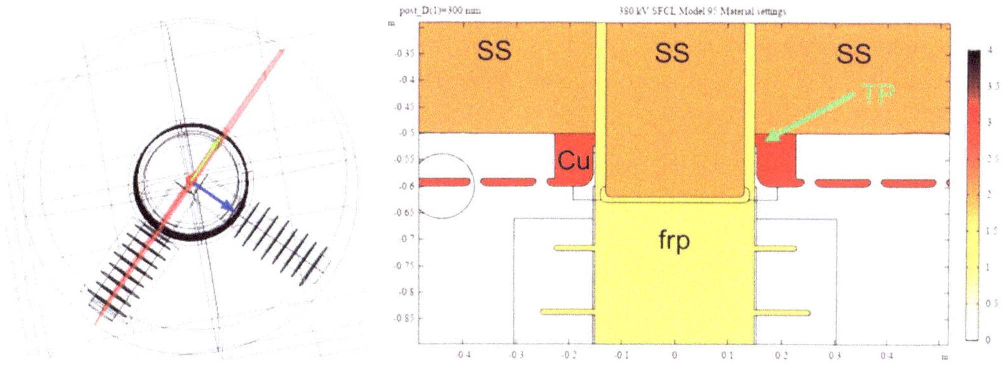

Figure 4.4.15: Left: Cut plane (red) through one leg, right: View of cut plane around the shield for the pole insulators. Material view at triple point. Material number 0 (LN_2, white), 1 (FRP, yellow), 2 and 3 (metal, orange and red). Liquid nitrogen gap at triple point TP is 4 mm

A parameter sweep to the inner rounding of the pole insulator shield delivered maximum field values below the limit of 5 kV/mm (see Table 4.4.2) for radii of 20 mm and 30 mm but not for 10 mm.

The higher permittivity of the solid caused a field enhancement for some locations in liquid nitrogen along the rod and ribs. Due to the higher background field, an exceeding of the limit of 2.7 kV/mm was calculated especially around the highest rib (Fig. 4.4.16).

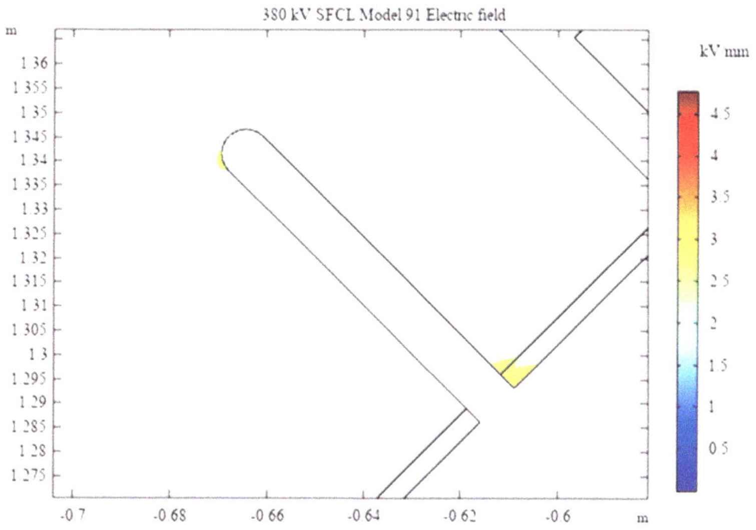

Figure 4.4.16: Field plot around top rib with minimum data range set to 2.7 kV/mm, i.e. lower values are not shown (white color). Field values in liquid nitrogen exceed the limit of 2.7 kV/mm at the transition zone between rod and rib and near the outer end of the rib.

A calculation for a material exchange from FRP to PE with ε_r = 2.4 for the highest rib was then calculated and it turned out that the field limit was still exceeded. The next trial was another sweep done with the permittivity values between 2.4 and 3.5 for the complete post insulator material including the ribs. A possible material choice may be polycarbonate with ε_r = 2.8 for 77 K which has more suitable material properties then PE. Nevertheless, even for permittivity of 2.4 the field limit of 2.7 kV/mm was exceeded in liquid nitrogen at the rod around the highest rib region.

In a further step the total liquid nitrogen volume, in which the electric field strength *E* exceeds a value of 2.75 kV/mm, was calculated for 4 legs and ε_r = 2.8 was calculated to 5 cm³. For *E* > 2.70 kV/mm it was calculated to be 11 cm³. For one pole insulator, the investigated space is shown in Fig. 4.4.17 as green and blue.

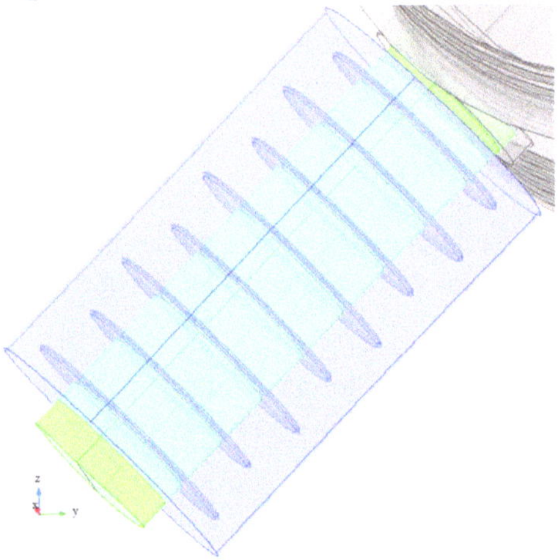

Figure 4.4.17: Liquid nitrogen volume around one pole insulator. Cylinders around the rod part of the solid insulator are green. Blue is the additional liquid nitrogen volume for the field calculation in order to also investigate the rib influence on the liquid nitrogen space located not directly at the rod part

Another calculation was done for the volume around a pole insulator but without a 4 mm thin liquid nitrogen sheet around the rod shaped part of the solid insulator (investigated space only blue in Fig. 4.4.17). The volume where the field value exceeded 2.70 kV/mm was lower than 0.3 cm³. This part is labelled as "Emax LN_2 around pod ribs" in Table 4.4.6 (B). Such a low volume with a maximum value of 2.9 kV/mm may be acceptable.

Presently, it cannot be judged if higher field limits in the liquid nitrogen space near the ribs than near the rods can be accepted.

The main values for the maximum field for the different configurations around the bipods are summarized in Table 4.4.6. The exceeding of the field strength limit of 2.7 kV/mm in the liquid nitrogen along the highest part of the leg is marked with "A" in Table 4.4.6 as an indication of a potential critical space. Verification of the bipod high voltage design by experiments will be necessary.

Table 4.4.6: Maximum electric field for different permittivities of the bipod insulator material

post_er i.e.: ε_r for post insulator material	E_{max} LN_2 along leg low	E_{max} LN_2 along leg middle	E_{max} LN_2 along leg high	E_{max} LN_2 triple point	E_{max} LN_2 around leg ribs	E_{max} LN_2 bipod shield fillet	E_{max} LN_2 shielding ring	E_{max} insulator at metal insert
	kV/mm	kV/mm	kV/mm	kV/mm	kV/mm	kV/mm	kV/mm	kV/mm
2.4	2.2	2.5	2.9	0.018	2.7	3.0	3.8	4.6
2.8	2.3	2.6	3.1	0.019	2.9	2.9	3.8	4.5
2.9	2.3	2.6	3.2	0.019	2.9	2.9	3.8	4.4
3.0	2.3	2.7	3.2	0.019	3.0	2.9	3.8	4.4
3.5	2.5	2.9	3.4	0.020	3.2	2.8	3.7	4.3
Target	2.7	2.7	2.7	< 0.1	$2.7 \leq \varepsilon_r \leq 3.8$	3.8	3.8	5.0
Status for $\varepsilon_r = 2.8$	OK	OK	A	OK	OK? B	OK	OK	OK

4.4.3.4 Cryostat Terminations

A termination similar to the 220 kV limiter installed in Moscow made from a few tori did not fulfil the field strength limit requirements. Therefore, the configurations shown in Fig. 4.4.18 were investigated. A spherical termination (Fig. 4.4.18 left), a termination made from two hemispheres and one cylinder (Fig. 4.4.18 middle) and an "extended termination" version (e.g. Fig. 4.4.18 right) made from one hemisphere connected by a cylinder to an object made of two spherical segments were examined. The bottom part of the bushing was located within a cut of the termination in order to save height.

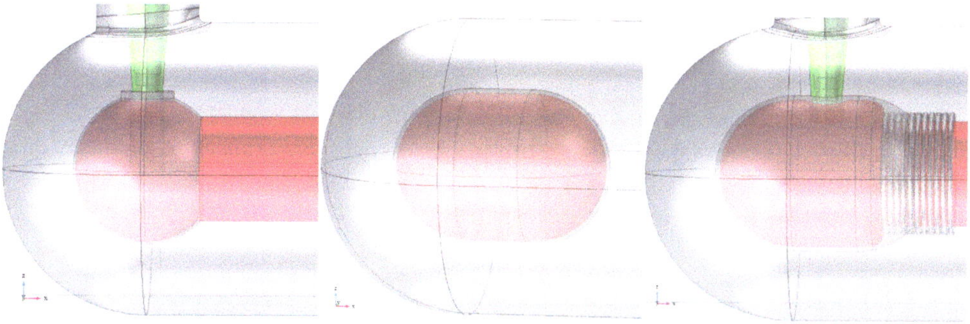

Figure 4.4.18: Investigated cryostat termination versions. Left: Spherical termination with common tube as simplification for shielding rings (model 73). Middle: Termination made from two hemispheres connected by a cylinder. No shielding rings are shown (model 76). Right: An extended termination made from one hemisphere connected by a cylinder to an object made of two spherical segments (model 112)

The spherical termination fulfilled the field strength requirements but was very sensitive against changes from the ideal shape. The termination made of two hemispheres connected by a cylinder fulfilled the field strength requirements, too, and allowed more deviation from the perfect shape but it requires more length.

A few "extended termination" models were investigated in order to reduce the cryostat length and fulfil the field limits simultaneously. In that context the field strength around the termination was not only investigated with regard to the termination shape but also the outer diameter of the shielding ring.

Field strengths of extended terminations are shown in Fig. 4.4.19. The fillet of the innermost part of the termination is important for keeping the field strength limit. In Fig. 4.4.19 left it can be seen that the innermost fillet of the termination (dashed red circle) allowed to shift shielding rings to be located nearer towards the bushing which helped to save space. In Fig. 4.4.19 right the fillet was increased for termcyl_dr = 13 mm, resulting in a closer distance to the outermost shielding ring but lower field strength on a critical space.

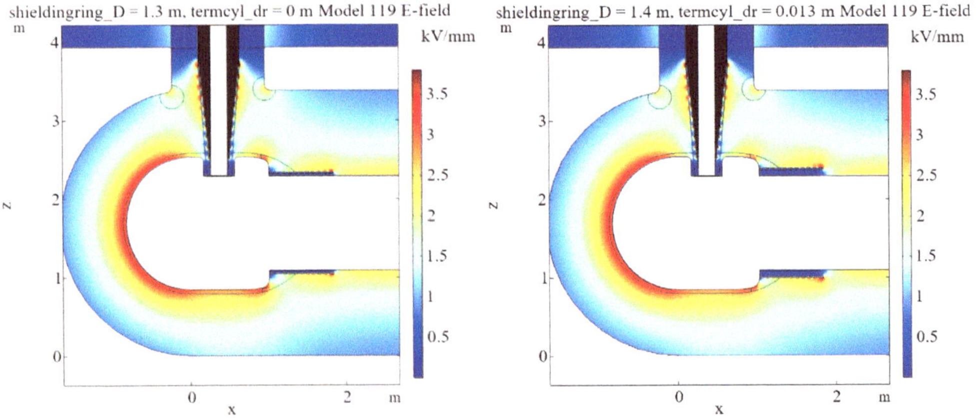

Figure 4.4.19: Left: Field strength of extended termination and shielding rings with a diameter of 1.3 m. The fillet in the dashed circle allowed a lower length of the SFCL. Right: Field strength of extended termination and shielding rings with a diameter of 1.4 m. The fillet in the dashed circle is termcyl_dr = 1.3 cm larger than the fillet in Fig. 4.4.20 left.

The variation of the two fillet radii and the two shielding ring diameters (shieldingring_D) delivers a number of four different shapes (model 119). Maximum field strengths between 4.0 kV/mm and 4.3 kV/mm were calculated with a Comsol parameter sweep. The high field volume calculation can be seen in Table 4.4.7.

Table 4.4.7: High field volume for different shielding ring diameters and versions

shieldingring_D	termcyl_dr	$V(E > 3.80$ kV/mm)	$V(E > 3.85$ kV/mm)	$V(E > 3.90$ kV/mm)
m	m	dm³	dm³	dm³
1.30	0.000	0.969	0.6369	0.4295
1.30	0.013	0.473	0.2575	0.0693
1.40	0.000	0.009	7.70E-4	7.37E-5
1.40	0.013	1.54E-4	0.0000	0.0000

According to the calculated high field volumes, the version with a shielding ring diameter of 1.4 m may still be acceptable. Another termination version was obtained by cutting a full inner hemisphere of the termination in the axial direction and shifting the outermost shielding rings in the newly formed liquid nitrogen space within the termination (Fig. 4.4.20). This resulted in a maximum field strength of 3.8 kV/mm.

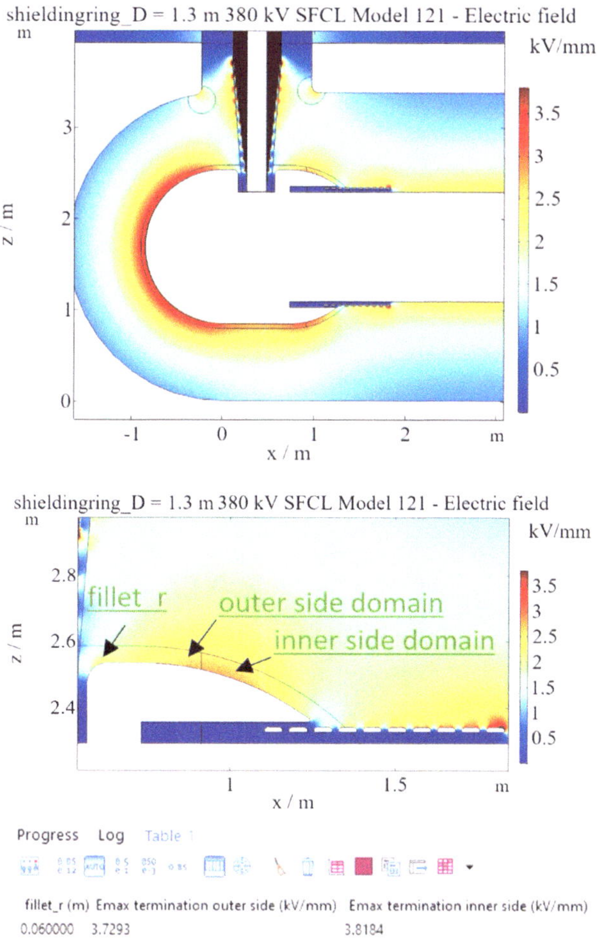

Figure 4.4.20: Top: Extended termination with axial cut in inner hemisphere in order to provide space for shielding rings. Bottom: Extended termination with axial cut in inner hemisphere in order to provide space for shielding rings. Zoom to the inner hemisphere and calculation of maximum fields

It can be concluded, that the field calculation delivers some solutions for different shapes, which can be considered as candidates for high voltage experimental verification.

4.4.3.5 Cryostat Dimension Estimation

The Comsol submodels for a more detailed calculation of the cryostat dimension are shown in Fig. 4.4.21. No single model could be built due to the model size being too large. Therefore, different parts were put together as shown in this figure.

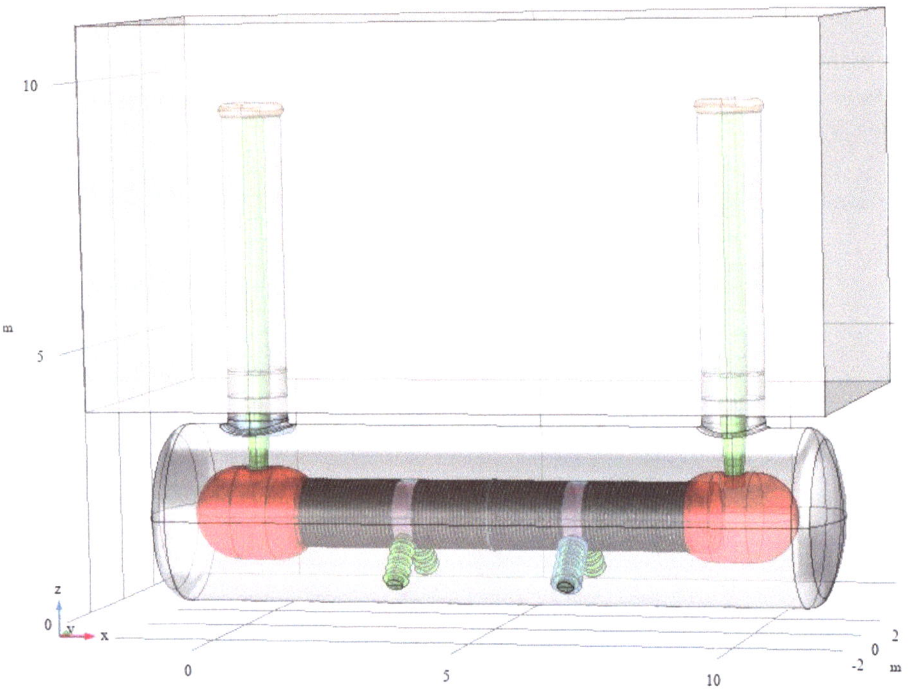

Figure 4.4.21: Comsol geometry model 131 representing one possible combination for dimension estimation

For this final geometry file, a SFCL with 93 modules was selected with extended terminations and torispherical heads for the cryostat. Only 91 shielding rings were used because each bipod shield can also incorporate one module. In addition, the total length was reduced by assuming that two shielding rings can be implemented in each inner termination hemisphere. The bipod shields were each increased to a length of 0.493 m, compared to the bipod mini model where only a length of 0.485 m was used. The additional length can help to make the bipod shield manufacturing and assembly less complicated. All this together with information from separate mini models, delivers the following data for the cryostat length (see Table 4.4.8).

Table 4.4.8: Length dimensions of the cryostat

Cryostat region	Data
Hemispherical heads of cryostat	2 · 3.38 m / 2 = 3.38 m
Cylinders with current lead connection	2 · 0.73 mm = 1.46 m
Inner spherical part of termination, cut	2 · 3.38 m / 4 / 3 = 0.56 m
Shielding rings with gaps	91 · 0.075 m + 0.05 m = 6.88 m
Bipod shields with module, with gap	2 · 0.505 m = 1.01 m (1.025 m)
With this data the following sum can be calculated	
Cylindrical part of cryostat for hemisphere head	9.905 m
Cylindrical part of cryostat for torispherical head	12.1 m
Total inner length of cryostat (head to head)	13.29 m (13.41 m)

For comparison, the data derived from the simplified illustrative model in Fig. 4.4.21 are given in brackets and are in good agreement with the data derived from detailed submodels.

The total height of the cryostat and the bushings are given in Table 4.4.9.

Table 4.4.9: Heights of the SFCL sections. The connection of the cryostat cylinder and a dome affects the position, depending on the heights in overlapping space.

Cryostat region	Data
Hemispherical heads of cryostat	3.38 m (because of 3D cut 3.27 m)
Dome height	1.10 m (cylinder length)
Bushing RT length (with fixation flange)	4.86 m
Height on top of bushing (without top flange and without top high voltage torus)	9.34 m (9.23 m)

It should be mentioned that the dimension values depend on several assumptions (e.g. type of termination). Experiments are recommended to gain a deeper understanding of dielectric strength under (fast) transient and high voltage application of the foreseen test voltages and materials. Such new information can have strong influence on further design optimization.

In addition, it must be considered that the electrical fields are calculated for cryogenic high voltage operation conditions. This means that, for design and manufacturing the room temperature dimension values require a material-specific increase compared to the values of the models in order to consider the shrinkage effect due to cooling.

4.4.4 High Voltage Bushings

The capacitor bushing design is based on experience with bushings from the company HSP in Cologne used at KIT with liquid nitrogen. The maximum voltages applied at these bushings under cryogenic conditions are 230 kV AC (rms), 325 DC and LI or LIC up to 365 kV. For positive polarity, 10 LI tests were performed up to 475 kV and 3 LIC tests were performed with 500 kV with one of the KIT bushings. The LIC tests were unintentionally caused by failure of the connected samples. In case of 500 kV tests the test object was a short cable sample for the superconducting 110 kV Siemens power transmission cable project. Although the voltages mentioned with KIT bushings are considerably lower than the required operation and test voltages for the foreseen 380 kV SFCL, the bushings are considered for the high voltage design estimations because there is only little knowledge of bushings for higher voltages which were successfully operated under cryogenic conditions.

An overview of the high voltage bushings used in our experiments is shown in Table 4.4.10. These bushings can be considered as suitable or almost suitable test bushings for superconducting components of the 110 kV voltage level (U_m = 123 kV) where the AC test voltage is 230 kV and LI test voltage is 550 kV.

Table 4.4.10: Properties of 3 different capacitor bushing types manufactured by HSP for KIT

HSP-Type	EK200spez 672.1780.070000.01	EK 230 spez 672.1480.51101.01	GEK 325 spez 601456
ITEP facility	HAIhKE, Fatelini1	Fatelini 2	Fatelini 2
Acceptance test	18.09.1997	04.02.2009	08.02.2017
U_{AC} / kV	200	230	230
U_{DC} / kV			325
U_{LI} / kV	550 (?)	550	550
U_{AC} withstand 1 min / kV	240		276
U_{LI} withstand / kV	±600	±605	+550 / -605 (design: 660?)
U_{LIC} withstand / kV		-696	±666 (design: 696?)
l_{total} / mm	2013	2355	2355
$l_{below\ flange}$ / mm	2013 – 1232 = 781	2355 – 1415 = 940	2355 – 1415 = 940
$l_{grading\ cryo}$ / mm	218	350	350
$l_{grading\ RT}$ / mm	1107	1185 (?)	?
d_{cryo} / mm	190	230	230
d_{inner} / mm	49 ... 51	70 (?) ... 72	70 (?) ... 72
m / kg	70	130	130
p / bar abs.	5	4	4

For a 380 kV bushing the upscaling is performed taking into consideration the HSP standard delivery program for U_m = 123 kV and U_m = 420 kV [HSP2021] in Fig. 4.4.22.

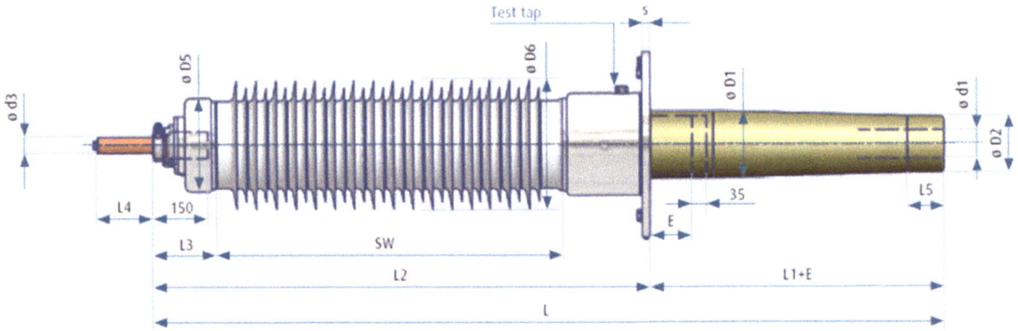

Figure 4.4.22: Dimensions on SETFt bushing of HSP [HSP2021]

Several other dimensions are derived from practical experience at KIT [Fink 1998] e.g. height for radiation plates. With this input, the following chapter estimates the main dimensions.

4.4.4.1 Bushing Length

A comparison with Fig. 4.4.23 is shown for a better understanding of the following selection of dimensions. Data from HSP bushing are taken from [HSP2021].

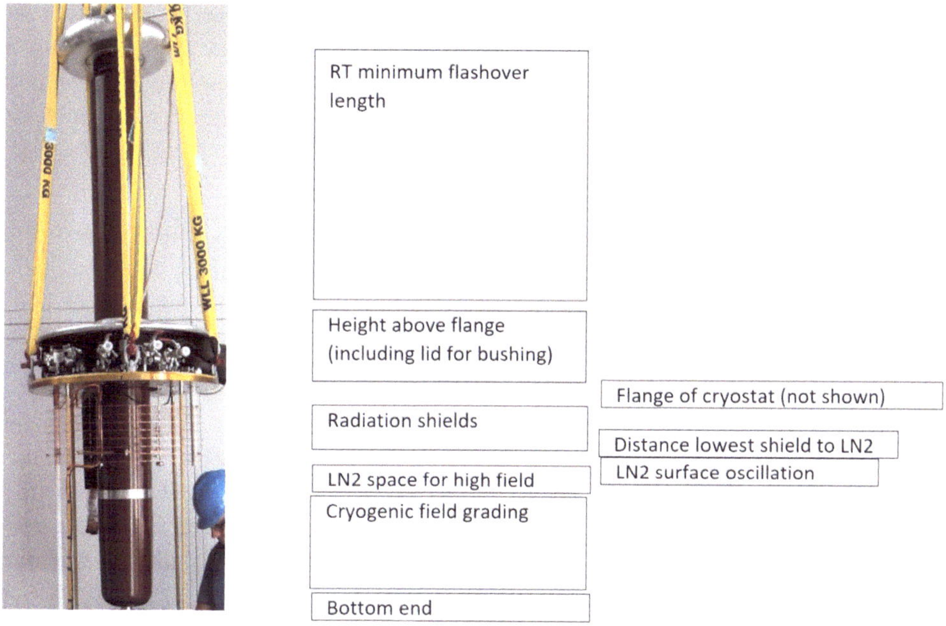

Figure 4.4.23: HSP DC bushing for 325 kV at KIT

The bushing is selected as an outdoor bushing. The room temperature part above the mounting flange of the bushing on the cryostat dome can therefore be derived from the HSP bushing SETFt 1550-420-B with its minimum flashover distance of 4354 mm. The room temperature length (above the fixation flange of the dome, including the lid thickness of 60 mm) is thus 4859 mm (derived from length L_2 in [HSP2021]). The length for thermal grading below fixation flange with radiation plates is 300 mm. It is not useful that the lowest radiation shield is submerged within liquid nitrogen. The distance of the liquid nitrogen level to the lowest radiation plate is chosen to be 100 mm.

The refill process may cause a certain oscillation of the liquid nitrogen level. In addition, oscillations caused by mechanical stress (e.g. earthquake) or a certain loss of liquid should be tolerated (unintentional activation of overpressure relief, tightness problems). Therefore, the selected liquid level oscillation range is 50 mm. The space directly above the cryogenic field grading length is expected to be exposed to a field strength where the absence of gas is preferred. This results in a liquid length above the field grading of 200 mm.

The dielectric strength of liquid nitrogen is often considered to be similar to the strength of transformer oil. In larger scale experiments scattering of the flashover voltage for liquid nitrogen is found which makes it necessary to increase the cryogenic (flashover) field grading length for liquid nitrogen in comparison to the oil part of a transformer bushing. Since no literature data are available for this issue at the needed voltage level, the ratio of the KIT test bushing for 230 kV to a 110 kV HSP bushing SETFt 550-123-B E600 is taken as 350 mm / 225 mm and multiplied with the field grading length 710 mm in oil for the 420 kV HSP bushing. In addition, a safety margin of 20% is added leading to a cryogenic field grading length of 1325 mm.

The aluminium layers used for field grading within the bushing cannot just be cut with direct contact at the bottom side to the surrounding liquid. The bottom length for an HSP bushing for 420 kV in oil is 95 mm. Hence, the bottom length below field grading is selected as 100 mm for the cryogenic bushing. With the values above the total length below cryostat flange is 2075 mm and the total length of the bushing is 6934 mm (around standard length for U_m = 550 kV at HSP).

4.4.4.2 Bushing Diameter

Due to the dimensions and the high current, a hollow cylinder is assumed as the first version for the current carrying part of the current lead that is located inside the bushing. A skin depth for copper of 9 mm was assumed for a rod or tube at room temperature. Considering a maximum current of 5 kA and a maximum current density of 1 A/mm^2 the required minimum copper inner tube diameter was calculated to be 168 mm and the outer tube diameter to be 186 mm.

An outer radius of 243 mm was calculated for a reduced skin depth to 3.3 mm under cryogenic conditions. Since such a large copper tube diameter cannot be accepted, segmentation or heating must ensure that a diameter calculated at room temperature is not exceeded. No risk of destruction is assumed for cryogenic temperatures by selection of the copper cross section according to room temperature skin depth because skin depth increases with temperature increase.

An inner diameter of the bushing of 210 mm was selected regarding a certain distance to the inner copper tube with its outer diameter of 186 mm.

An outer bushing diameter of D = 530 mm was designated in comparison to several HSP standard bushings and bushings used at lower voltages for cryogenic application at KIT. Since these values scatter in a wide range, this issue must be discussed intensively with the bushing manufacturer.

4.4.4.3 Field Grading Strategy and FEM Models

Ribs along the cryogenic field grading flashover lengths may be discussed because such ribs are reported to increase the flashover voltage [Koo2017]. A fine grading is required with a 1 mm up to 3 mm distance between layers for capacitor bushings [Küchler2018] resulting in a number of layers in the range of (530 mm – 210 mm) / 2 / 2 mm = 80. For the FEM model, a layer number of only 11 was selected in order to reduce the calculation effort. Different grading modes will be possible, depending on priority of field grading [Philippow1982].

For the FEM modeling of the 380 kV bushing a simple grading was selected. Firstly, starting for each layer length with the part corresponding to the length of the outermost (i.e. grounded) layer, a linear increase of each of the room temperature layer lengths, and a linear increase of each of the cryo layer lengths, by considering the length of the innermost layer. The representation of the bushing in the FEM model is shown in Fig. 4.4.24.

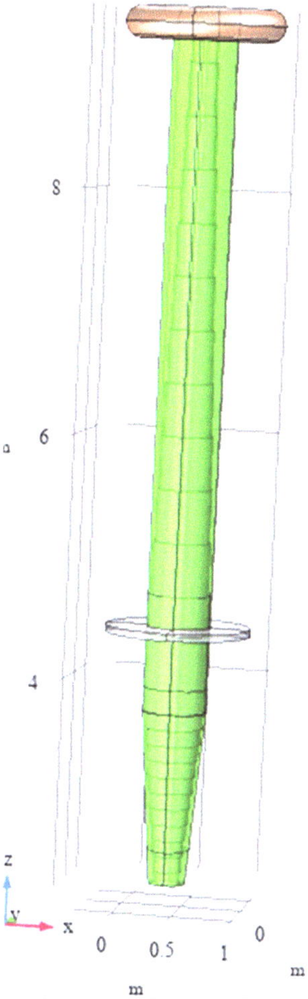

Figure 4.4.24: Bushing within Comsol FEM model. The room temperature length above the lid is 4799 mm. The lid thickness is 60 mm. Distance from lid to lowest liquid nitrogen level is 450 mm. Ground layer length of in vessel side is 650 mm and cryogenic field grading length is 1325 mm. Total bushing length below lid is 2075 mm.

4.4.5 Voltage Distribution inside Limiter

Due to the large spatial expansion of the current limiter element (see Fig. 4.4.1) and resulting parasitic effects, in general an asymmetrical voltage distribution is expected during transient failure stresses like lightning impulse, chopped lightning impulse or switching impulse. The parasitic effects are caused by the residual inductance of the bifilar wound HTS tape arrangement as well as the capacitive coupling between the HTS tapes and the stray capacitance between the entire limiting element and the earthed cryostat. This section considers the resulting voltage distribution over the entire limiting element (and respectively, the modules). As with the 115 kV current limiter in the SuperLimiter project [Schmidt2014], the simplified capacitive chain network model is used, as well as an extended RCL network model analysis with the inclusion of all estimable electrical quantities. Finally, the field strength within the bifilar wound insulation system inside the coils is determined by means of the finite element method in order to optimize the insulation arrangement.

4.4.5.1 Calculation of the Electrical Quantities

Before the simulation can be carried out using a capacitive chain network model and the more detailed RCL network model, the individual parameters and electrical quantities must be determined. For this, the detailed material properties and geometric dimensions are required with which the electrical quantities resulting from the selected arrangement can be determined. The current limiter consists of 85 (case A) or 93 (case B) HTS modules connected in series. The modules in turn consist of 6 parallel, bifilar wound coils. Each bifilar coil in turn consists of two parallel-wound tape-conductors insulated from each other by a layered insulating arrangement (Fig. 4.4.25).

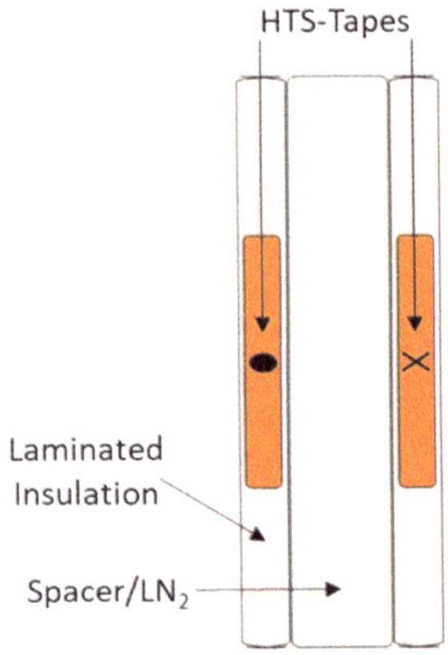

Figure 4.4.25: Exemplary front view onto the sectional build-up of the layered insulating system of one bifilar wound tape conductor arrangement

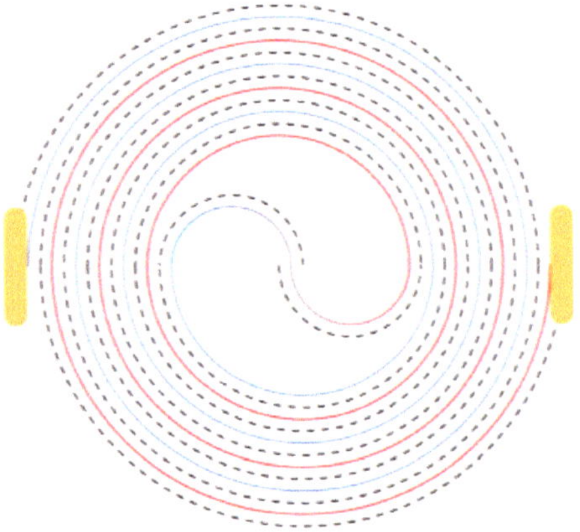

Figure 4.4.26: Exemplary representation of one bifilar coil. The red and blue line show the bifilar wound tape conductors

The electrical equivalent circuit (EEC) of one bifilar coil (Fig. 4.4.28) consists in superconducting mode at normal operation of a parallel connection of a mutual inductance L_{Tape} and a coupling capacitance C_{Coupling}. After the transition from superconducting to normal conducting (quench) due to a fault, a resistance R_{HTS}, caused by the commutation of the current from the HTS layer to the normal conductive silver layer, is added in series to the inductance L_{Tape} (Fig. 4.4.28). Accordingly, the electrical equivalent circuit of one module consists of a p_{L}-fold parallel connection of the electrical equivalent circuit of one bifilar coil. To this the stray capacitance C_{Earth} per module must be added, which acts between each module and the earthed cryostat. Finally, N modules connected in series from Fig. 4.4.27 result in the electrical equivalent circuit of the entire limiting element.

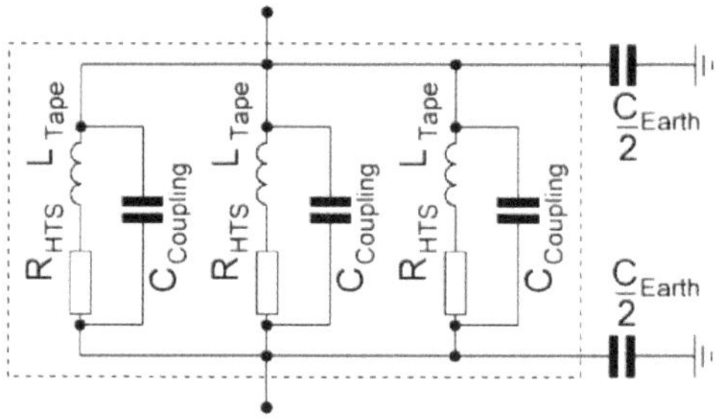

Figure 4.4.27: Electrical equivalent circuit of one module consisting of several electrical equivalent circuits of bifilar coils

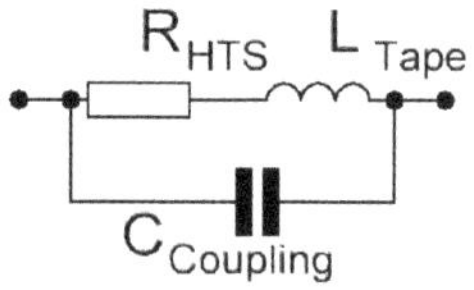

Figure 4.4.28: Electrical equivalent circuit of one bifilar coil (Fig. 4.4.26) inside a module

In Table 4.4.11, the necessary parameters and values for calculating the electrical quantities for the selected configurations from Table 4.3.2 and Table 4.3.3 are shown. The material selection and their dimensions are based on experimental research results [Bauer2016, Gromoll2020, CIGRE2021, Kreuz2022] as well as the current state of the art for the construction of bifilar wound coils and their insulation arrangement [Schacherer2009, Tixador2018, CIGRE2021].

Table 4.4.11: Material properties and assumptions for geometrical dimensions for calculating the electrical parameters

Property	Case A	Case B
HTS tape dimension A_{HTS}	12 mm x 105 µm	
Tape edge radius r_{edge}	25 µm	
Tape arrangement	TWIN	
Distance twin tape to twin tape d_{tt}	2 mm	
Kind of tape insulation	Laminated	
Insulation material	Polyimide	
Insulation thickness d_{ISO}	130 µm	
Insulation permittivity $\varepsilon_{r,ISO}$	3.14	
TWIN-tape thickness with insulation	470 µm	
TWIN-tape height with lamination	20 mm	
Spacer material	Polyimide	
Spacer thickness d_{Spacer}	300 µm	
Spacer permittivity $\varepsilon_{r,Spacer}$	3.14	
LN_2 gap d_{LN2}	1.44 mm	
LN_2 permittivity $\varepsilon_{r,LN2}$	1.43	
Tape length of single coil l_{HTS}	76.6 m	62.4 m
Parallel bifilar coils p_L	6	
Inner module diameter d_i	200 mm	
Height of module	25 mm	
Number of modules in series connection N	85	93
Outer radius of shielding rings $r_{a,S}$	750 mm	
Shielding ring width (each module) l_{Shield}	30 mm	
Inner radius of cryostat $r_{i,C}$	1500 mm	

Coupling Capacity C_{Coupling}

The coupling capacitance is the capacitance between the parallel tape conductors of a coil that are separated by an insulation system. The arrangement corresponds approximately to that of a plate capacitor whose capacitance can be expressed by the following general equation.

$$C = \frac{\varepsilon_0 A}{\sum_{i=1}^{N} \frac{d_i}{\varepsilon_{ri}}} = \frac{\varepsilon_0 A}{\frac{d_1}{\varepsilon_{r1}} + \frac{d_2}{\varepsilon_{r2}} + \cdots + \frac{d_N}{\varepsilon_{rN}}} \qquad (4.4.3)$$

Here the dielectric of the plate capacitor is made of N layers with a thickness d_i and a relative permittivity ε_{ri}. The thicknesses d_{ISO}, d_{Spacer} and d_{LN2} as well as the permittivities $\varepsilon_{r,ISO}$, $\varepsilon_{r,Spacer}$ and $\varepsilon_{r,LN2}$ are used to calculate the coupling capacitance C_{Coupling}.

Earth Capacity C_{Earth}

The earth capacitance is the capacitance between each module and the earthed cryostat. Neglecting the ends, the arrangement corresponds to a coaxial capacitor whose capacitance is calculated with the general equation.

$$C_{\mathrm{Earth}} = \frac{2\pi\varepsilon_0\varepsilon_{LN2} l_{Shield}}{\ln\left(\frac{r_{a,S}}{r_{i,C}}\right)} \qquad (4.4.4)$$

For the calculation each module inside the limiter element is assigned to a shielding ring with a width of l_{Shield} = 30 mm. The radius $r_{a,S}$ corresponds to the inner radius of the cryostat and $r_{i,C}$ to the outer radius of the shielding ring.

Stray Inductance L_{Tape}

In addition to the self-inductance of the tapes, the non-ideal coupling between the windings of the bifilar coil lead to small stray inductances. The coupling factor *k* itself cannot be calculated, but was determined experimentally in an equivalent arrangement to be roughly 0.97.

To determine the self-inductance of a bifilar wound tape, the following empirical equation from [Grover2018] was applied.

$$L_s = 0.001 n^2 a P f \qquad (4.4.5)$$

Here, a corresponds to the mean radius of the coil and n to the number of turns. The factors P and f are geometric correction factors whose values can be taken from the tables 1 and 3 on page 477 and 481 in [Grover2018]. To determine the stray inductance L_{Tape}, the calculated self inductance L_s must be multiplied by the factor$(1 - k^2)$.

$$L_{Tape} = (1 - k^2) \cdot L_s \qquad (4.4.6)$$

Tape Resistance R_{HTS}

The tape resistance, which occurs during quench operation, depends on the HTS type used and knowledge about their composition and thermal and electrical properties. This allows the temperature-dependent progression of the specific resistance to be calculated, as can be seen in Fig. 4.4.29 and Fig. 4.4.30 as example for the HTS tape from the companies SuperOx and THEVA.

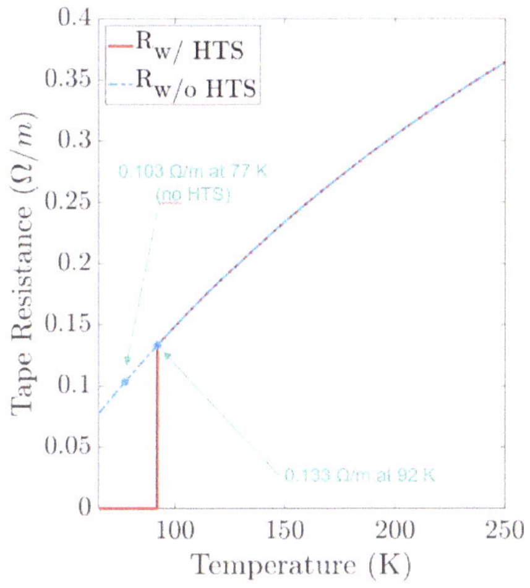

Figure 4.4.29: Temperature-dependant specific tape resistance during quench operation for a SuperOX type (KIT data)

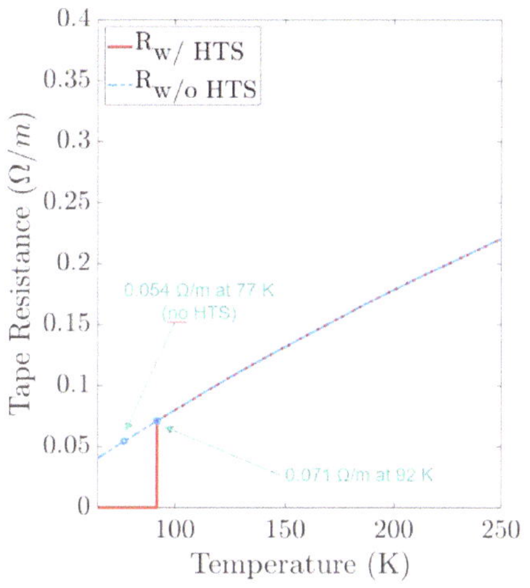

Figure 4.4.30: Temperature-dependant specific tape resistance during quench operation for a THEVA type (KIT data)

Determined Electrical Quantities

Table 4.4.12 shows the electrical values determined for the two application cases A and B using the Eq. 4.4.3 to 4.4.6 and the values from Table 4.4.10.

Table 4.4.12: Resulting electrical quantities for the application cases A and B (compare Fig. 4.4.34)

Property	Case A	Case B
Stray inductance L_{Tape}	21.5 µH	20.4 µH
Inductive coupling factor k	0.97	
Coupling capacitance $C_{Coupling}$	3.43 nF	2.79 nF
Tape resistance of one bifilar coil at 250 K R_{HTS}	13.8 Ω	11.2 Ω
Earth capacitance C_{Earth}	3.44 pF	
Copper connector resistance R_{Con} (assumed)	0.1 µΩ	
Copper connector inductance L_{Con} (assumed)	0.1 µH	

4.4.5.2 Voltage Distribution according to Capacitive Chain Network

In [Schmidt2014], for the analysis of the asymmetrical voltage distribution, it was assumed that the capacitive coupling and earth components mainly determine the voltage distribution at high frequencies. This results in the simplified equivalent circuit in Fig. 4.4.31 for the current limiter element.

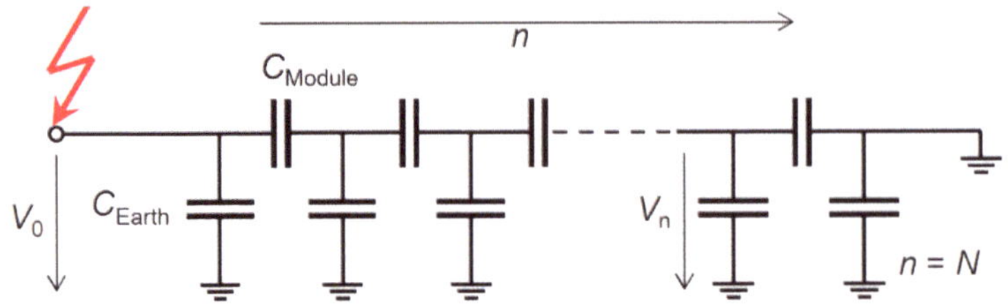

Figure 4.4.31: Simplified capacitive EEC according to [Schmidt2014]

Equation 4.4.7 determines the voltage distribution for the capacitive chain network model. Here N is the total number of modules in series and n is the number of the respective module existing of $p_L = 6$ bifilar coils in parallel (see Table 4.4.11).

$$V_n = V_0 \frac{\sinh(N - n)\alpha}{\sinh N\,\alpha} \qquad (4.4.7)$$

$$\alpha = \sqrt{\frac{C_{Earth}}{C_{Module}}} \quad \text{with} \quad C_{Module} = p_L \cdot C_{Coupling} \qquad (4.4.8)$$

Using Eq. 4.4.7 and 4.4.8 and the capacitance values from Table 4.4.12, the values in Table 4.4.13 and the curves in the diagrams in Fig. 4.4.32 and Fig. 4.4.33 are obtained. For better clarification, the relative relationship between the theoretical linear stress distribution and the calculated asymmetric stress distribution is shown. The highest possible voltage amplitude of the chopped lightning impulse with 1570 kV is used as the input voltage.

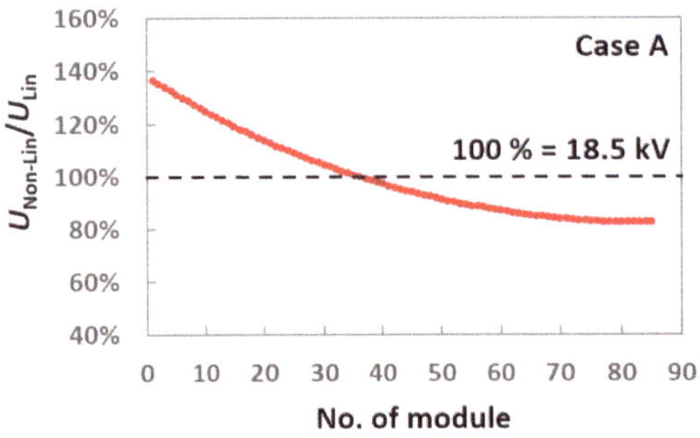

Figure 4.4.32: Course of the asymmetric voltage distribution according to the capacitive chain network model for case A

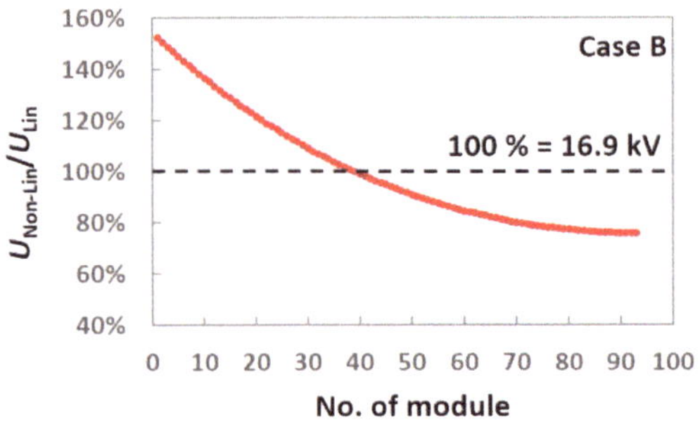

Figure 4.4.33: Course of the asymmetric voltage distribution according to the capacitive chain network model for case B

Table 4.4.13: Results of the asymmetric voltage distribution according to the capacitive chain network model for case A and B

Property	Case A	Case B
Coefficient α	0.01293	0.01433
Max. amplitude @ LIC	1570 kV	
Max. linear voltage per module	18.5 kV	16.9 kV
Max. voltage of module @ LIC	25.3 kV	25.7 kV
Max. relative voltage ratio	137 %	152 %

The results from Fig. 4.4.32 and Fig. 4.4.33 as well as Table 4.4.13 show the expected asymmetrical voltage distribution. The voltages across the first modules are always the highest, with a voltage of 25.3 kV in case A and 25.7 kV in case B. Despite the different electrical quantities, the absolute overvoltage is almost the same. For case B, however, this means a significantly larger relative voltage increase of 152 % (compared with 137 % in case A).

To verify the stresses, the values here were compared with those from [Schmidt2014]. For the coils, there was a calculated linear voltage stress of 31 kV. Based on the calculation of the asymmetric stress using a capacitive chain network model, a maximum voltage of 51 kV was calculated. The maximum relative voltage is 162 %, which is not very far away from case B considered here. However, the absolute voltage values are almost twice as high as those here. Therefore, this first estimation assumes that the voltage increases in case A and B are manageable. However, this depends on the resulting electric field strength in the HTS tape winding insulation (see section 4.4.5.4).

4.4.5.3 Voltage Distribution according to RLC Network

Since the capacitive chain network model considers only the acting capacitances, additional simulations were carried out with a more developed RCL network model. Using this model, not only the asymmetric voltage distribution was simulated when a transient overvoltage occurs in normal operation, but also when the current limiter is in quench operation. In addition, estimated electrical parameters were assigned to the connections between the coils, as these are made of normal copper (Table 4.4.12).

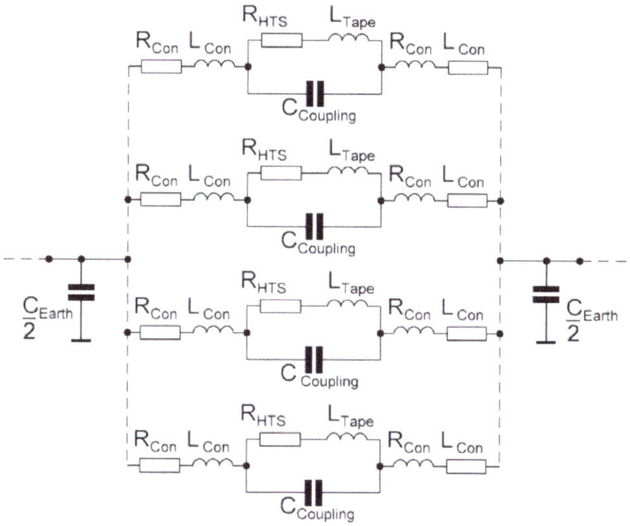

Figure 4.4.34: Example of an RCL sub-network consisting of four parallel bifilar coils.

Furthermore, a more representative voltage stress could be considered. By means of the Test Data Generator (TDG) from IEC 60076-6, both a standardised chopped lightning impulse voltage curve (Fig. 4.4.35) with a rise time of 1.2 µs and a chopping time of 6 µs and a standardised switching impulse voltage curve (Fig. 4.4.36) with a rise time of 250 µs and a tail time of 2500 µs were considered. Although there is no definition of the cut-off time except that it must be as short as possible, it is responsible for the transient response due to its much higher slope and is therefore absolutely worth knowing. Therefore, this was evaluated from the standardised chopped lightning impulse voltage from IEC 60076-6 and is approximately 200 ns.

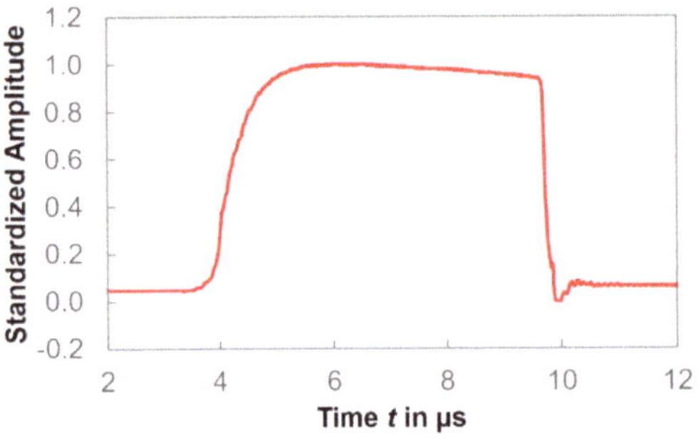

Figure 4.4.35: Course of the standardised chopped lightning impulse voltage curve from IEC 60076-6

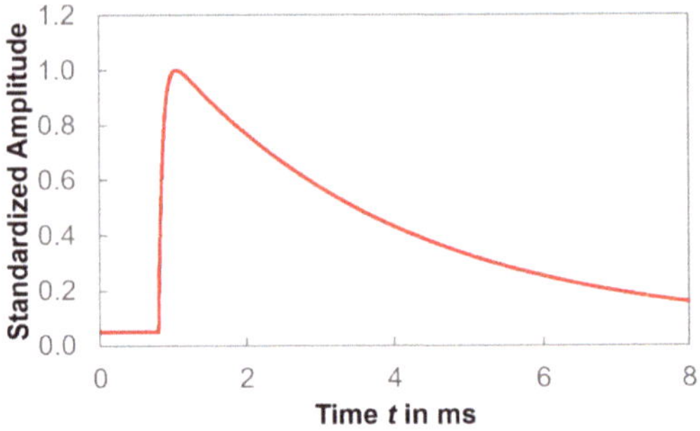

Figure 4.4.36: Course of the standardised switching impulse voltage curve from IEC 60076-6

With the values from Table 4.4.12 assigned to the RCL network model, the asymmetric voltage distribution for the stress with a chopped lightning impulse voltage could be simulated. In Fig. 4.4.37 and 4.4.38 the courses of the voltage drop for case A and B are shown. The voltage drop corresponds to the difference over every eighth module measured in the simulation. In addition, the highest emerging absolute and relative results are shown in Table 4.4.14.

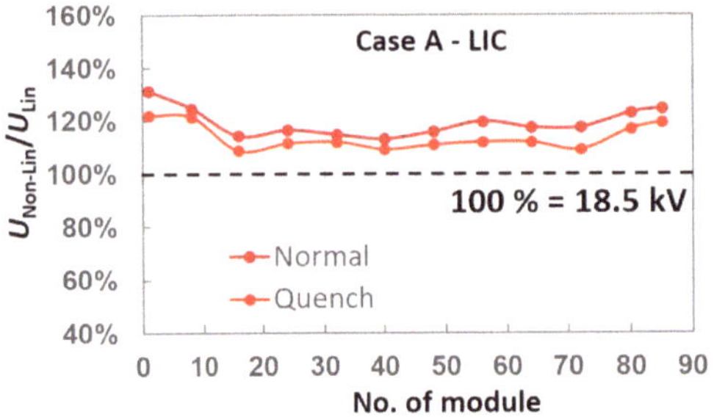

Figure 4.4.37: Course of the asymmetric voltage distribution according to the RCL network model for case A und LIC stress

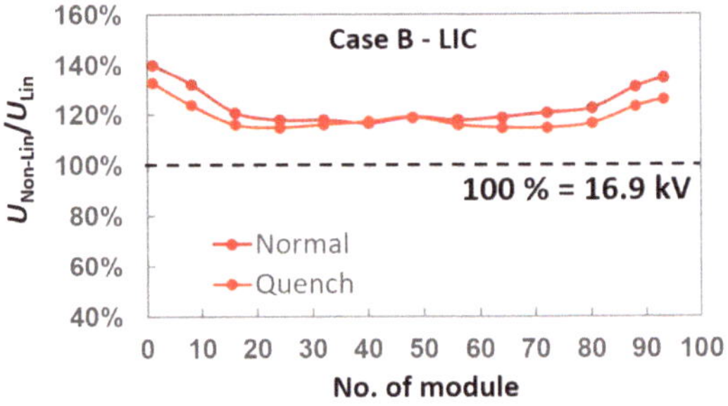

Figure 4.4.38: Course of the asymmetric voltage distribution according to the RCL network model for case B under LIC stress

Table 4.4.14: Results of the asymmetric voltage distribution according to the RCL network model for case A and B

Property	Case A - LIC		Case B - LIC	
Operation	Normal	Quench	Normal	Quench
Max. amplitude @ LIC	1570 kV			
Max. linear voltage per module	18.5 kV		16.9 kV	
Max. voltage of module @ LIC	24.2 kV	22.5 kV	23.7 kV	22.5 kV
Max. relative voltage ratio	131 %	122 %	140 %	133 %

The curves in Fig. 4.4.37 and 4.4.38 do not show a steady decrease in voltage with increasing module number. Instead, the voltage seems to increase and decrease periodically. It is noticeable that the individual simulated voltage curves are overlaid with higher-frequency components, which is due to the oscillating circuit behaviour of the electrical components R, C and L. These oscillations lead to constructive and destructive interferences whereby maxima and minima occur in the overvoltage.

When comparing the absolute values between quench and normal operation, it can be seen that the maximum voltage rise is lower with quench. This is mainly due to the fact that the additional resistance has a damping effect on the amplitude.

Comparing the absolute and relative results with those from section 4.4.5.2, Table 4.4.13, the maximum values of the RCL network model with 24.2 kV (131 %) in case A and 23.6 kV (140 %) in case B are slightly lower but in the same order of magnitude. Again, as before, it must be pointed out that these simulated results only reflect a trend. However, the fact that the values here and those from section 4.4.5.2 are of the same order of magnitude and lower compared to those from [Schmidt2014] again leads to the assumption that the voltage stresses here are manageable.

For the same electrical configuration, the asymmetrical voltage distribution for the stress with a switching impulse voltage results in the curve in Fig. 4.4.39 for case B.

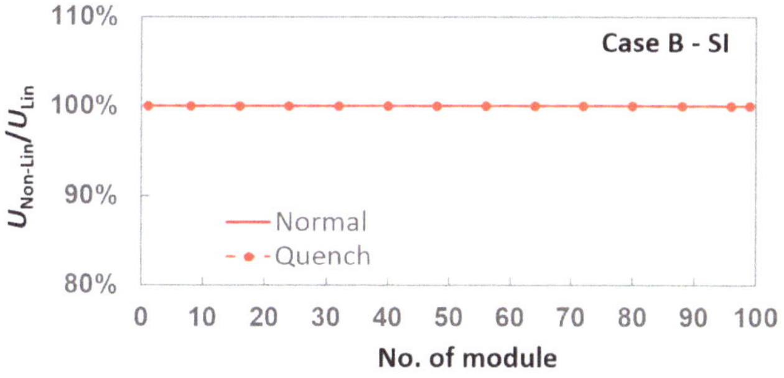

Figure 4.4.39: Course of the asymmetric voltage distribution according to the RCL network model for case B under SI stress

There is no asymmetrical stress distribution in the case of a switching impulse voltage during normal or quench operation. This is due to the fact that the maximum voltage slope of 250 µs, which is a few orders of magnitude slower than for the LI or LIC, is too low to cause an asymmetric voltage distribution.

4.4.5.4 Electrical Field Strength inside Winding

After the maximum voltage drop per coil has been estimated using the results from sections 4.4.5.2 and 4.4.5.3, the electric field strength in the winding insulation system can be calculated and the insulation system can be designed. For this purpose, the plane-parallel part of the bifilar winding, excluding the connection areas, including the insulation system, were considered by means of a 2D-FEM simulation and the electric field strength in the different insulation layers and media were analysed. The geometric and material properties correspond, as before, to the data in Table 4.4.11 and are shown in Fig. 4.4.40.

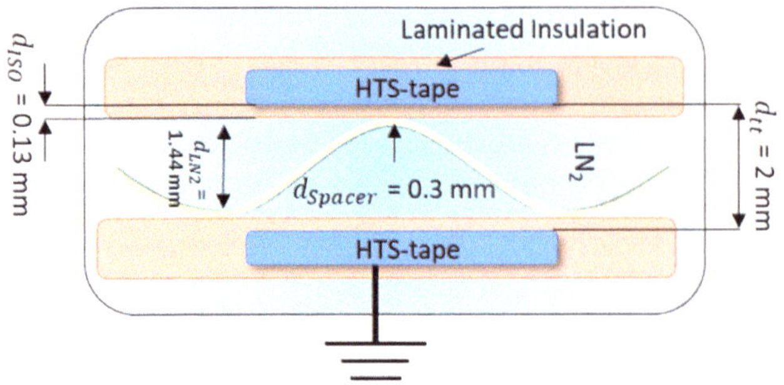

Figure 4.4.40: Example of the insulation system of a bifilar wound HTS tape winding

For the simulation and analysis of the field strength stress in the arrangement, additional boundary conditions had to be defined. The wave-shaped spacer was considered to be flat, resulting in a layered arrangement. Due to the undefined edge of an HTS tape, which leads to the problem of singularity in the simulation resulting in an incorrect calculation of the field strength, an edge radius of 25 µm was assumed, as shown in Table 4.4.11.

In addition, the peak field strength was averaged in the area of integration line 1 (Fig. 4.4.41), orthogonal to the tape, so that it can be better compared with the empirical investigated averaged peak field strengths. In the spacer and liquid nitrogen (or later gaseous nitrogen), the maximum field strength is determined in the approximately constant range in the middle of the arrangement (integration line 2 in Fig. 4.4.41). For the arrangement shown in Fig. 4.4.41, the resulting field strength values are summarized in Table 4.4.15.

In general, only the field strength stress in case B will be considered in the following, since the voltage stress is higher during normal operation and quench operation compared to case A.

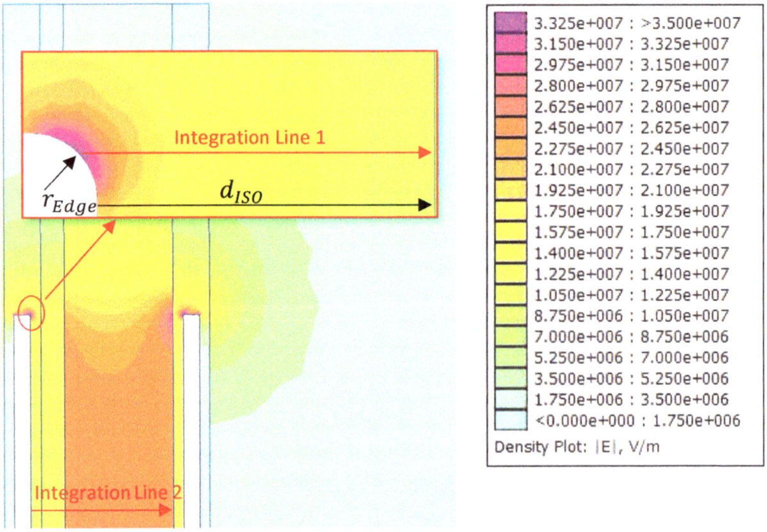

Figure 4.4.41: Electrical field distribution inside the winding arrangement of case B in normal operation at a maximum resulting voltage stress of 24.2 kV during LIC

Table 4.4.15: Resulting field strength stresses in the different insulating layers at 24.2 kV at normal operation during LIC

Property	Polyimide lamination	Polyimide Spacer	LN_2 gap
Field strength $\hat{E}$	15.9 kV/mm	6.5 kV/mm	15.3 kV/mm

For the operation under quench, the determination of the maximum field strength is more complex. In this case, it must be assumed that the superconductor heats up to 360 K, which means that some or all of the liquid nitrogen will evaporate to gaseous nitrogen (N_2).

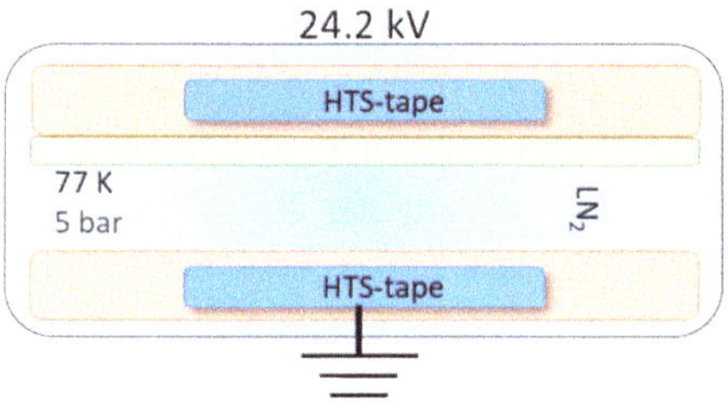

Figure 4.4.42: Exemplary representation of the plane layered arrangement under LIC stress during normal operation

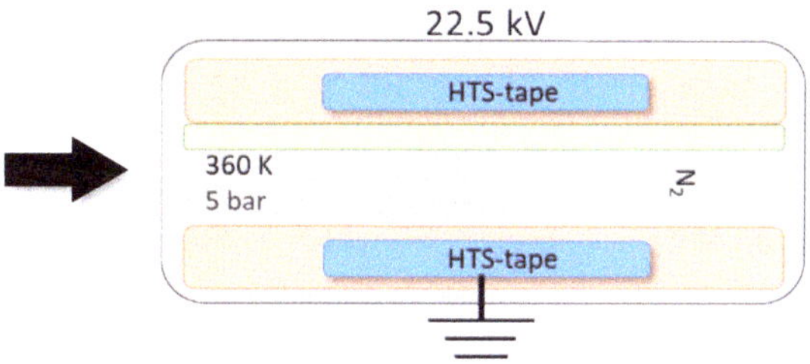

Figure 4.4.43: Exemplary representation of the plane layered arrangement under LIC stress in quench operation where the LN_2 is evaporated to N_2

Table 4.4.16: Resulting field strength stresses in the different insulating layers at 22.5 kV at quench operation during LIC.

Property	Polyimide lamination	Polyimide Spacer	N_2 gap
Field strength $\hat{E}$	12.0 kV/mm	0.5 kV/mm	13.9 kV/mm

The evaporation of the LN_2 (Fig. 4.4.42) to N_2 (Fig. 4.4.43) leads to a considerably lower electrical strength of the N_2 gas gap, which can be calculated by means of Paschen's law. The calculation results in about 13.9 kV/mm for the strength of N_2 at the ambient conditions of 5 bar and 360 K. Compared with the electric field strength stress from Table 4.4.16 Table 4.4.16, it becomes clear that this stress can lead to pre-discharges in the N_2 gap (see Fig. 4.4.44).

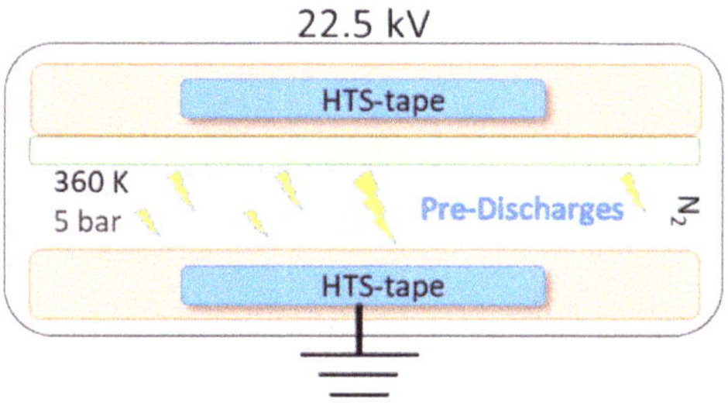

Figure 4.4.44: Exemplary representation of the plane layered arrangement under LIC stress in quench operation where the LN_2 is evaporated to N_2 and pre-discharges occur

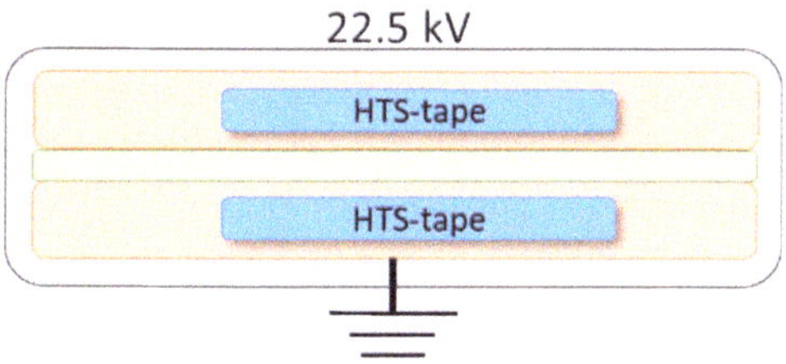

Figure 4.4.45: Exemplary representation of the plane layered arrangement under LIC stress in quench operation where no N_2 gap is present from the electrically point of view

Ultimately, the pre-discharges in the N_2 gap mean that this area is not available for electrical insulation. Thus, the polyimide lamination and the polyimide spacer have to withstand the entire stress with a lower electrical insulation distance (Fig. 4.4.45).

Table 4.4.17: Resulting field strength stresses in the different insulating layers at 22.5 kV at quench operation during LIC where no N_2 gap exists from electrically point of view, caused by pre-discharges

Property	Polyimide lamination	Polyimide spacer
Field strength $\hat{E}$	48.9 kV/mm	40 kV/mm

The values in Table 4.4.17 for the electric field strength in the two remaining polyimide layers show an approximate increase of up to four or five times. In [Kreuz2022] investigations were carried out regarding the lightning impulse withstand voltage of tape insulation arrangements in liquid nitrogen. An average strength value of approximately 190 kV/mm was determined for a 130 µm thick polyimide film, which is almost four to five times higher than the expected field strength stress here. Considering that LIC is a rare and not a permanent exposure, for which a much higher safety factor would be necessary, the factor of four to five times is considered sufficient.

However, these are theoretically determined values that have not yet been confirmed by experimental studies. Therefore, further experimental investigations regarding the electrical strength of such compact bifilar-wound insulating assemblies are necessary in order to be able to make a precise statement.

4.4.5.5 Summary

The results from the capacitive chain and RCL network simulations show comparable results and allow a representative classification with comparison to an existing system. Therefore, the absolute asymmetric voltage drops across the modules and coils are considered manageable. The electric field strength stress in the winding insulation, resulting from the simulated expected LIC voltage stress, seems feasible when compared with the experimental findings regarding the electric strength of solid insulations between tape arrangements in liquid nitrogen under impulse stresses.

In general, the simulated results of the asymmetrical voltage distribution are to be understood as an estimate and are subject to an unknown variance in their absolute value. In addition, the simulation methods do not consider superpositions from running time effects as they can arise from travelling waves. In reality, for example, the electrical quantities do not occur as lumped but extended components. In addition, not all parasitic influences can be considered in the simulation

This also applies to the design of the winding insulation. For such a complex arrangement as the one considered here, no experimental electrical strength tests have yet been undertaken with regard to the breakdown or flashover characteristics under AC, LI, LIC and SI stress and under the complex environmental conditions, such as pressure or gas bubble formation. However, these experiments are necessary in order to design the winding insulation in such coils reliably.

4.4.6 Travelling Wave Voltages

Voltage and current changes propagate along electrical lines or in electrical equipment at the speed of light c in the corresponding medium. This guided electromagnetic wave is called a travelling wave in electrical power engineering. In general, the influences of travelling waves must be taken into account if the propagation time τ of the travelling wave in the spatial area of length x under consideration is not negligible compared to the time Δt in which the voltage changes. For a voltage error lower than 0.5 %, travelling waves have to be considered in the following case [Küchler2018]:

$$\tau = \frac{x}{c} > \frac{\Delta t}{15} \quad \Leftrightarrow \quad x > \frac{\Delta t}{15} \cdot c \qquad (4.4.9)$$

In the case of a transient LIC pulse the decisive change time of the voltage is the cut-off time of 200 ns (section 4.4.5.3). Assuming a medium light velocity of $c = c_0/\sqrt{\varepsilon_r} = 200$ m/µs considering the relative permittivity of the high voltage insulation, travelling waves should be considered if the length x exceeds a value of about 3 m, which is clearly the case for a SFCL considering the entire tape length in series connection.

Due to propagation time delays and reflection processes, travelling waves can lead to significantly increased voltage stresses in extended electric equipment which is well known e.g. in the field of transformers. To what extent this is the case in SFCL with bifilar wound coils is not clear. In general, travelling waves can propagate in the coils in two ways (Fig. 4.4.46):

- The transient voltage propagates in the circumferential direction into the centre of the bifilar coil (blue arrow) and from there back to the outer area (red arrow). Due to the long length of the superconducting tape in one bifilar coil, the voltage difference $\Delta V = V_{\text{out}} - V_{\text{in}}$ between output and input would be very large.
- Due to the capacitive coupling of the windings of the bifilar coil, it is also possible that the transient voltage propagates in radial direction through the coil (violet arrow). In this case a significantly smaller voltage difference ΔV is expected.

Due to the design of bifilar coils, the voltage difference ΔV is applied directly between two adjacent windings and stresses the winding insulation in the outer area. To investigate the impact of travelling waves on bifilar coils more in detail, a simulation model was developed and verified with the help of measurements on real coils and modules.

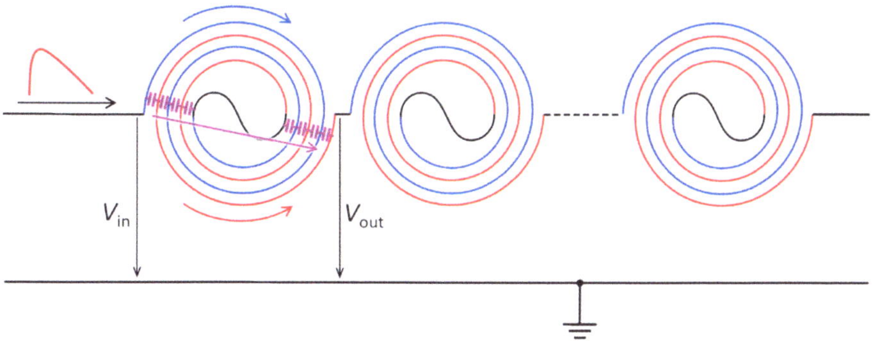

Figure 4.4.46: Propagation ways of travelling waves in bifilar coils of a SFCL

4.4.6.1 Verification of Simulation Model with Measurements

The SFCL module under investigation (Fig. 4.4.47, Table 4.4.18) exists of five parallel bifilar coils with an outer diameter of 400 mm, which are connected with a common inner contact of a diameter of 145 mm and fixed on a plate of glass fibre reinforced plastic (GRP). The coils are wound with copper tape conductors so that an investigation at room temperature is possible. All other properties and dimensions correspond to those of the modules used in the project ECCOFLOW [Noe2012, Hobl2013]. In the frame of this project a medium voltage SFCL was developed with a rated voltage of 24 kV and a rated current of 1 kA.

Despite the significantly different boundary conditions of the SFCL, the construction of the module is not significantly different from the construction proposed here for the 380 kV SFCL. The total tape length of 15 m per single coil is about one fifth of the module for case A (Table 4.3.2) and the turn-to-turn distance of 1.3 mm is not significantly different from the 2 mm used here. As spacer material, the ECCOFLOW modules use a DMD (Dacron/Mylar/Dacron) material consisting of two outer resin-impregnated polyester fleeces and a central polyester film. Because of the limited dielectric strength of this material, a polyimide spacer is proposed in the case of the 380 kV SFCL (Table 4.4.11).

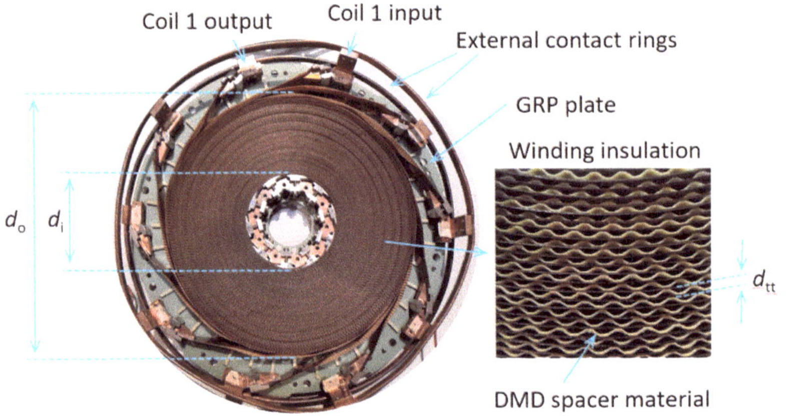

Figure 4.4.47: Investigated module with five bifilar coils (only coil 1 is labelled)

Table 4.4.18: Properties of the investigated module

Inner diameter d_i	145 mm
Outer diameter d_o	400 mm
Winding thickness $(d_o - d_i)/2$	127.5 mm
Dimension of copper tapes	0.1 mm x 12 mm
Tape length in single coil	$2 \cdot 7.5$ m = 15 m
Parallel windings p_L	5
Turn-to-turn distance d_{tt}	1.3 mm
Turn-to-turn winding insulation Thickness of Kapton layers Thickness of DMD spacer Thickness of air gap	 $2 \cdot 0.135$ mm 0.300 mm 0.730 mm

Three ECCOFLOW modules are available and connected in series (Fig. 4.4.48). At each module, the inputs and the outputs of the five bifilar coils are each connected with external contact rings with an outer diameter of 550 mm. The series connection is realized by connecting all outputs with the inputs of the next module. To replicate the influence of the earthed cryostat, a copper sheet ring with a diameter of 700 mm surrounds the three modules in series. The copper ring is connected to earth, which is the reference potential for the impressed and measured voltages.

Figure 4.4.48: Experimental setup to measure the voltage distribution in a series connection of three modules surrounded by an earthed copper ring to replicate the cryostat

The impressed impulse voltage is a burst impulse with a rise time of 5.1 ns and an impulse duration of 58.7 ns (Fig. 4.4.49), generated by an EFT 500 burst generator from EM TEST, connected to the input of module 1. A burst pulse is used to better resolve the expected short propagation times during the measurement. Furthermore, the reproducible generation of chopped lightning impulses (LIC), which are expected to have the highest impact, is not possible with the available equipment.

Voltage measurements were performed at the following positions:

- At the input of module 1 in parallel to the connected burst generator
- Between two adjacent input and output windings of module 1
- At the outputs of module 1 and module 3

All used probes and the oscilloscope have a minimum bandwidth of 500 MHz, which is necessary to resolve the burst impulses.

One single module is recreated as a 3D simulation model using the finite element method HFSS (High Frequency Structure Simulator) from Ansys (Fig. 4.4.50). The five bifilar coils with the same dimensions as in the experimental setup, the contact rings, the GRP plate (green) and the earthed copper ring (blue) are modelled. Due to limited memory resources, the details of the insulation materials between the windings are neglected and replaced by an area of a homogenous dielectric material. For the same reason the clamp connection at the inputs and outputs and at the centre connector are also simplified.

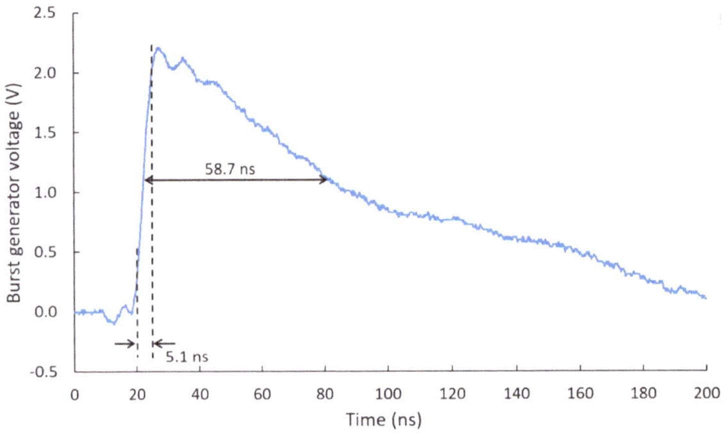

Figure 4.4.49: Generated burst impulse measured directly at the output of the burst generator without load

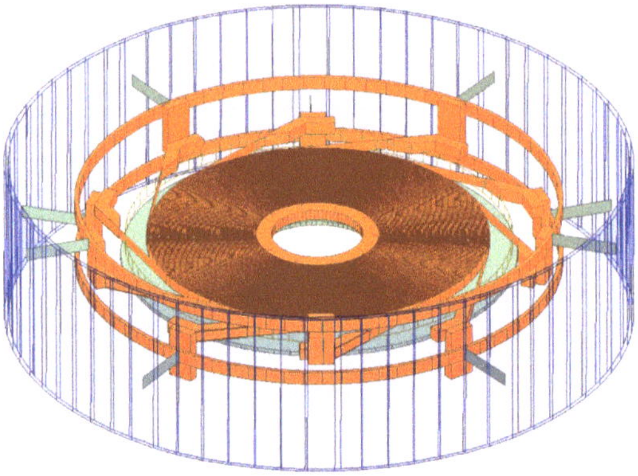

Figure 4.4.50: Ansys HFSS model of one module with surrounding earthed ring

Firstly, the model is simulated in Ansys HFSS in the frequency domain calculating the S parameter matrix. After the simulation in Ansys HFSS, the complete model is described by its S parameters for the frequency range at the terminals and can be implemented as an electronic component in Circuit Design from Ansys for simulations in the time domain. To simulate three modules in series all outputs of one module are connected with the inputs of the next module (see Fig. 4.4.51).

Two HFSS models are needed to simulate the modules in series correctly. The model for module 1 and 2 has only one contact ring at the inputs, because the contact ring at the inputs of the next module simultaneously takes over the function of the contact ring at the outputs of the previous module. Therefore, in this case only the model for module 3 has contact rings at inputs and outputs. Like in the experiment, the output of module 3 is open, which means that the termination impedance is infinitely large.

The measured burst impulse of the EFT 500 burst generator is applied via the internal resistance of the generator, R_{EFT} = 50 Ω, to the input of coil 1 of the first module. Corresponding to the experimental setup four voltage signals are recorded during the simulation at the input of module 1 (V_1), at the outputs of module 1 (V_2) and module 3 (V_3) and between the adjacent windings (V_4). For this, an extra terminal has to be defined in the HFSS model.

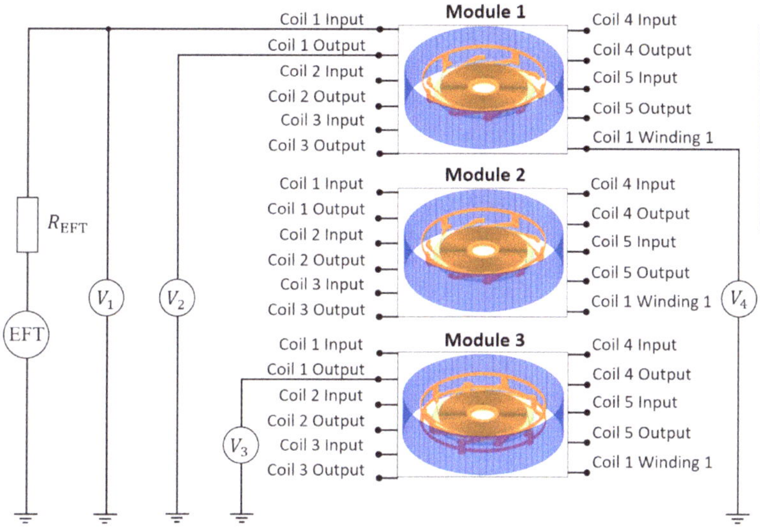

Connected	
Module n	Module n+1
Coil 1 Output	Coil 1 Input
Coil 2 Output	Coil 2 Input
Coil 3 Output	Coil 3 Input
Coil 4 Output	Coil 4 Input
Coil 5 Output	Coil 5 Input

Figure 4.4.51: Ansys Circuit Design model of three modules in series (connections not shown but given in the table)

Measured and simulated voltages at the inputs and outputs of the modules are shown in Fig. 4.4.52, referred to the peak value V_{max} of the input voltage at module 1. In principle, there is good agreement between simulation and experiment. The only value that was adjusted in the simulation to the measured values was the relative permittivity of the homogeneous medium in between the bifilar coils. This resulted in a value of ε_r = 1.69, which seems reasonable with regard to the structure of the winding insulation and the insulating materials used (Table 4.4.18).

In principle, three different travelling wave effects can be identified (Fig. 4.4.52):

- Time delays become visible at the front of the transient voltage, after which the transient voltage rises at the respective modules. Between the increase at the input of the first module and the increase of the voltage at the output of the third module, a time delay of about 4 ns was measured and of about 6 ns simulated. Assuming a medium light speed $c = c_0/\sqrt{\varepsilon_r} = 230\ \text{m/µs} = 230\ \text{mm/ns}$, this time delays correspond to propagation lengths of about 920 mm and 1380 mm in total or 307 mm and 460 mm per module, respectively. Comparing the winding thickness of the modules of 127.5 mm (Table 4.4.18) makes clear that the transient pulse is capacitively coupled or transmitted at the modules. The connection lines and the centre contact also have an influence on the propagation time, whose simplifications in the model can be the reason for the difference between measurement and simulation.
- The measured and simulated curve at the input of module 1 show approximately a doubling of the voltage, which classically occurs in the case of a reflection at the open end (impedance infinite). It seems that the pulse is reflected at the end of module 3 and propagates capacitively coupled back to the input of module 1.

- In addition to some oscillations in the course of the pulses, larger distortions can be seen in the range of 90 ns and 160 ns, especially of the voltage at the output of module 3. Measured from the start of the burst pulse at about 20 ns, this correspond to time delays of about 70 ns and 140 ns or propagation lengths of 16 m and 32 m. Sixteen meters is roughly the tape length of 15 meters in a single coil of the modules (Table 4.4.18), which leads to the conclusion that these distortions are caused by the part of the burst pulse following the windings in the circumferential direction. After the first 70 ns, this pulse part reaches the output of module 1 and couples then capacitively to the output of module 3. After another 70 ns the influence of the pulse running in circumferential direction along the windings through module 1 and 2 might be visible.

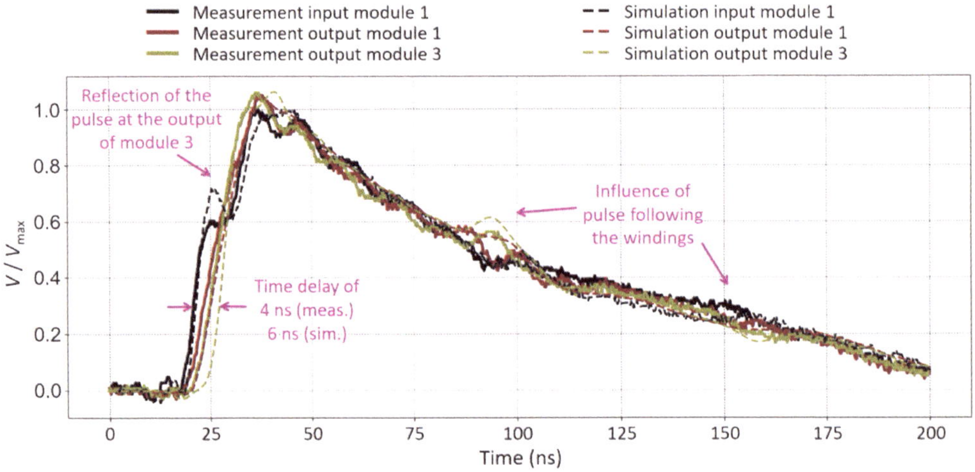

Figure 4.4.52: Comparison between measurement and simulation results of the input and output voltages

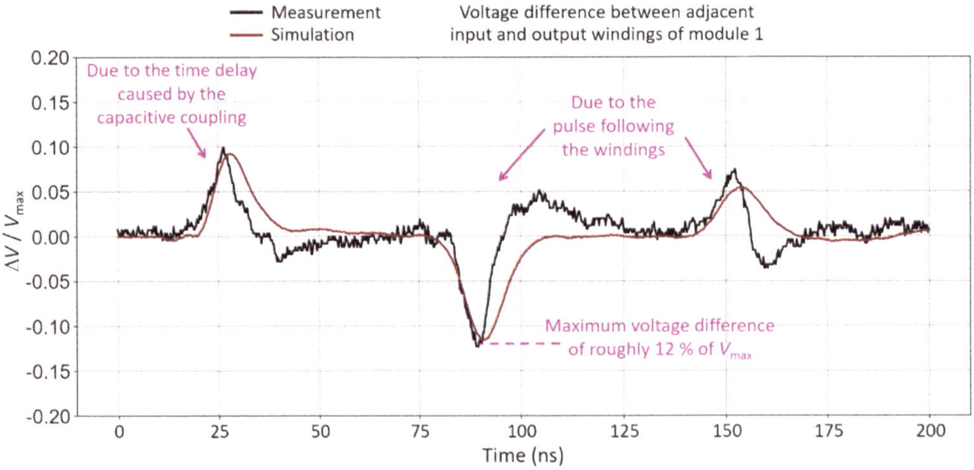

Figure 4.4.53: Comparison between measurement and simulation results of the voltage difference between adjacent input and output windings of module 1

The different travelling wave effects lead to corresponding voltage stresses on the windings of the modules. Measurement and simulation show that at the beginning of the burst pulse there is a voltage difference between adjacent input and output windings of module 1 of about 10 % of the peak voltage due to the time delay caused by the capacitive coupling (Fig. 4.4.53). At around 90 ns and 160 ns, additional maxima of the voltage stress between the windings occur caused by the propagation of the pulses along the windings. Here a maximum voltage difference of 12 % of the peak value of the applied voltage arises.

Overall, the agreement between simulation and measurement is very good. The measurement is disturbed by some overshoots and undershoots. Nevertheless, the time position and the amplitude of the overvoltages agree very well. In the following, further simulation results with the existing model of the module are shown, which cannot be verified with experimental investigations.

4.4.6.2 Influence of Modules in Series and Wave Impedance

To investigate the influence of more modules in series and the influence of the wave impedance of the connected lines terminating the output and the input of the SFCL, a model of the ECCOFLOW SFCL was developed by switching 12 modules in series. Additionally, impedances of $Z = 250\ \Omega$ parallel to the stack of modules were implemented (Fig. 4.4.54) to simulate the wave impedance of the connected lines of the grid. The model was excited again with a burst pulse in the first step.

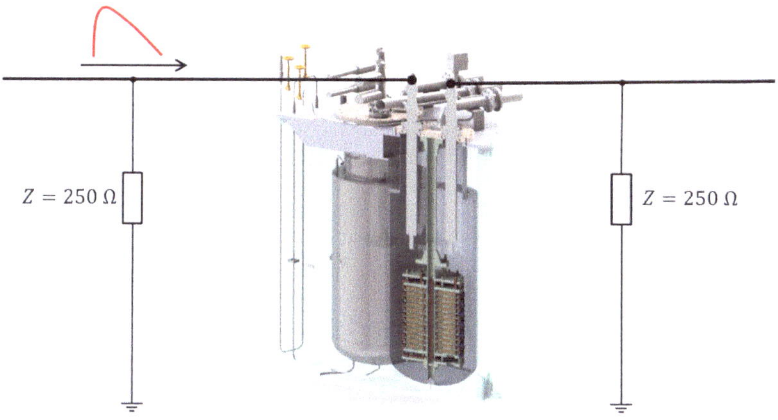

Figure 4.4.54: Simulated arrangement of the ECCOFLOW SFCL with 12 modules in series with an assumed wave impedance of 250 Ω of the connected grid, picture from [Hobl2013], modified

The results in Fig. 4.4.55 reveal that due to the simulated wave impedance of the grid the reflection at the output of the last module no longer occurs. There is a time difference of about 1.7 ns between the voltage rises at adjacent modules due to the capacitive transmission, whereby the amplitude of the first voltage peak decreases with module number. Additionally, a clear influence of the pulse components, which develops along the windings, is visible in the time ranges around 90 ns and 160 ns.

The voltage differences between adjacent input and output windings is affected in a similar way as in the simple model of the three modules with a maximum at the beginning of the burst pulse and further maxima of the voltage stress at 90 ns and 160 ns (Fig. 4.4.56). The maximum voltage difference of about 10 % of the peak value of the applied voltage generated at module 1 is in the same order of magnitude. The voltage maxima last longer, maybe due to the termination with the wave impedance.

It is clear, that the overvoltage at the winding insulation decreases with the module number. Thus, only the modules located at the beginning or end of an SFCL are highly stressed. Further investigations have shown that the overvoltage at the winding insulation within an individual module decreases towards the centre of the module.

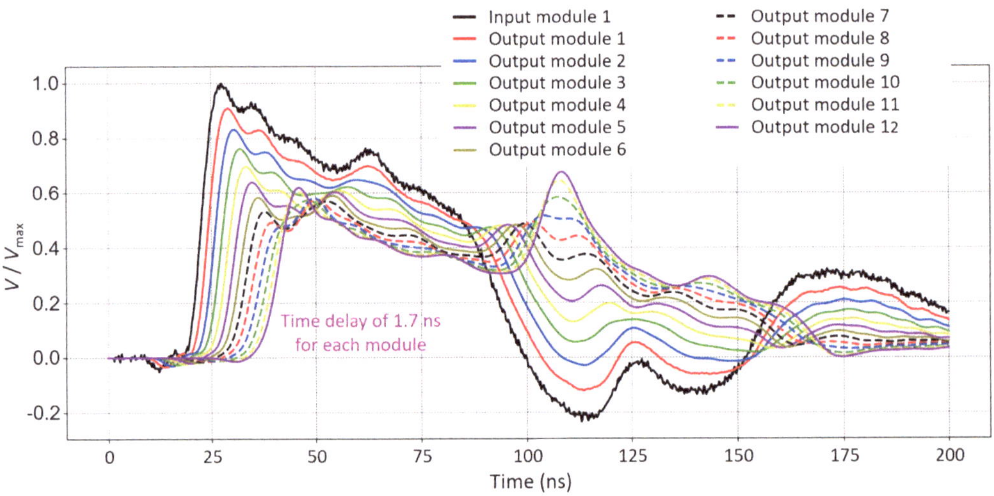

Figure 4.4.55: Simulated input and output voltages at the 12 modules in series with burst pulse

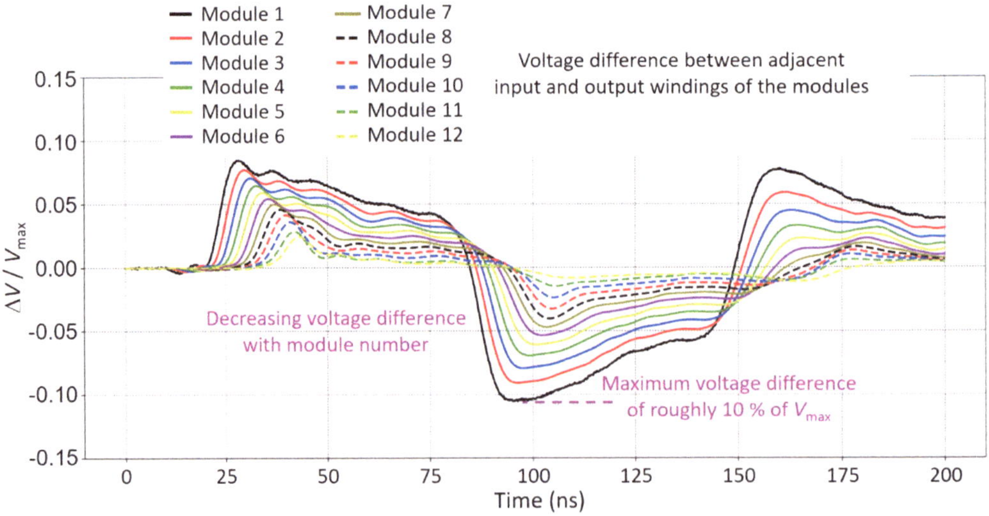

Figure 4.4.56: Simulated voltage difference between adjacent input and output windings of the 12 modules in series with burst pulse

Simulations with different wave impedances of the connected lines of the grid have shown that the maximum voltage difference at module 1 is only slightly dependent on the impedance (Fig. 4.4.57). Between 100 Ω and 400 Ω the voltage difference is in the range of 10.0 % and 10.6 % of the peak value of the applied voltage. Only with 50 Ω is the voltage difference slightly smaller than 10 %. Typical high voltage overhead lines have a wave impedance in the area of 300 Ω, power cables in the area of 40 Ω [Küchler2018]

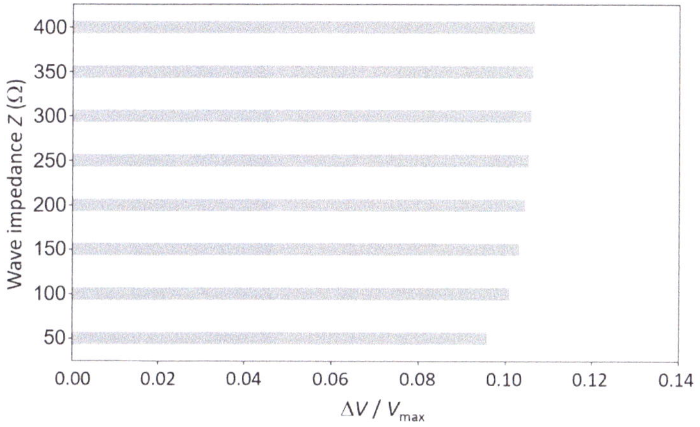

Figure 4.4.57: Maximum voltage difference between adjacent input and output windings of module 1 in dependence of the wave impedance of the connected lines

4.4.6.3 Simulation with LIC Pulse

To calculate the voltage distribution for the 380 kV SFCL influenced by travelling waves, modules with dimensions given in Table 4.3.2 (for case A) or Table 4.3.3 (for case B) have to be modelled and simulated in Ansys HFSS. Then 85 of these modules have to be connected in series and simulated in Ansys Circuit Design using a chopped lightning impulse (LIC) as applied voltage. At present, especially the first step, clearly exceeds the available computing capacity, so that an approximate solution is used here.

The tape length in a single bifilar coil for case A is 76.6 m, which is about five times the tape length of 15 m used in the modelled ECCOFLOW module. To approximate the impact of an LIC pulse, the existing ECCOFLOW model with 12 modules in series is used and simulated with a modified LIC pulse, which time base is divided by five, called LIC/5 in the following (Fig. 4.4.58). This results in a rise time of 240 ns instead of 1.2 µs and a cut-off time of 40 ns instead of 200 ns.

The results of the simulation show voltages of the different modules at the front of the LIC/5 and at the cut-off oscillations of the output. Due to the high voltage steepness especially at the cut-off, oscillations with peak values of more than 50 % of the applied voltage are generated (Fig. 4.4.59). A differentiation between the influences of the capacitive coupling of the travelling waves and the propagation along the windings is no longer possible.

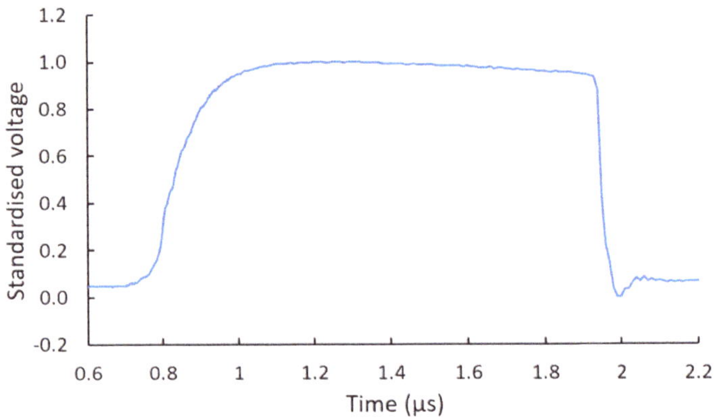

Figure 4.4.58: LIC/5 - LIC pulse with reduced time base divided by five

Despite the significantly longer time constants of the LIC/5 pulse compared with the burst pulse, the amplitudes of the voltage difference between adjacent windings of the modules are comparable (Fig. 4.4.60). At the cut-off of the LIC/5 pulse, a maximum voltage difference of again 10 % of the peak value of the applied voltage is generated. At the front, the voltage difference is only a maximum of 4 % at module 1. Again, the overvoltage at the winding insulation decreases with the module number.

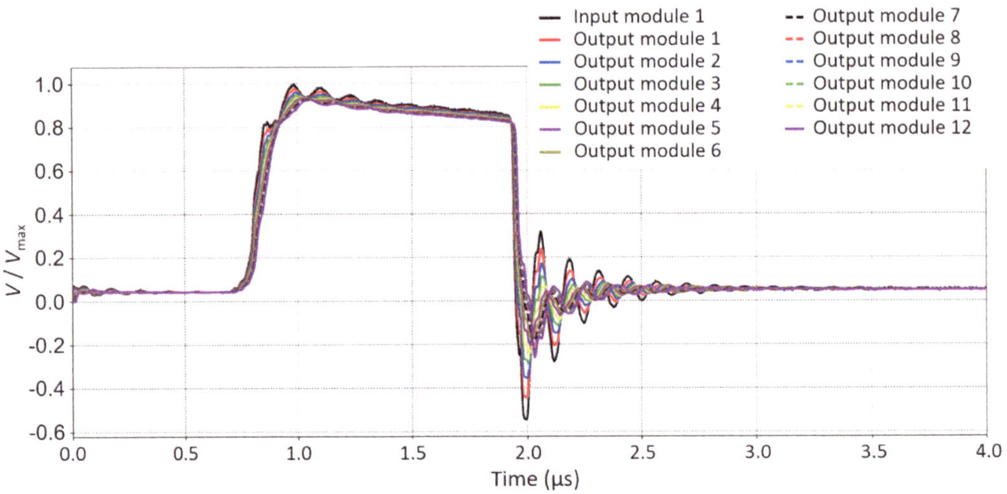

Figure 4.4.59: Simulated input and output voltages at the 12 modules in series with LIC/5 pulse

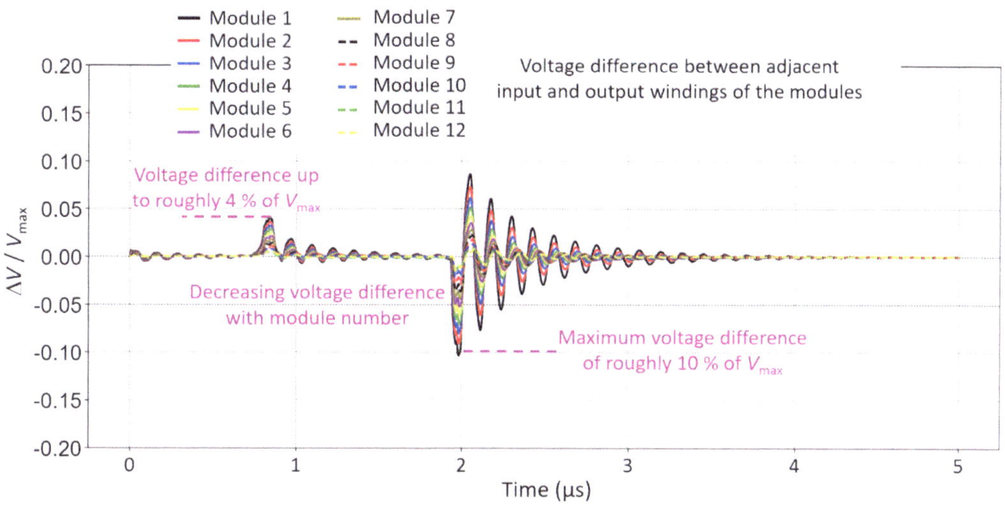

Figure 4.4.60: Simulated voltage difference between adjacent input and output windings of the 12 modules in series with LIC/5 pulse.

4.4.6.4 Summary

The measurement and simulation of travelling waves in bifilar coils of SFCL reveal clearly, that transient impulses propagate in both possible ways through the module. On the one hand the pulses propagate in the radial direction capacitively coupled through the module, on the other hand, second parts of the pulses run in the circumferential direction along the windings. Both parts have to be considered and lead to an increased voltage stress over the module and thus the winding insulation.

The estimation of the overvoltage in the case of an LIC pulse reveals a maximum voltage stress on the winding insulation of 10 % of the peak voltage (1570 kV) of the applied chopped lightning impulse voltage (Table 2.2.2). This means that the winding insulation will be stressed by roughly 160 kV instead of 24 kV, which was calculated using an RLC network (Table 4.4.14). But it has to be considered that the 160 kV overvoltage is an oscillating voltage of about 8 MHz, which means that the pulses have only a length of about 60 ns. Additionally, the voltage stress decreases with increasing module number and within each module from winding to winding. This indicates that only the modules at the beginning and end of the SFCL and the outer windings in particular require a reinforced winding insulation.

In any case, the current model describes the arrangement of the 380 kV SFCL only approximately. Further investigations are necessary to create a model of the planned large modules. In addition to the significantly longer tape length, the larger distance to the earthed cryostat wall could significantly change the behaviour when excited with transient pulses. In the same way, the simulation of more modules in series will significantly influence the behaviour of the SFCL system.

4.5 Current Limiting Reactor

As already mentioned in the previous chapters, a current limiting reactor (CLR) is used in addition to the fault current limiter. The current limiting reactor is connected in parallel to the current limiter (Fig. 4.5.1).

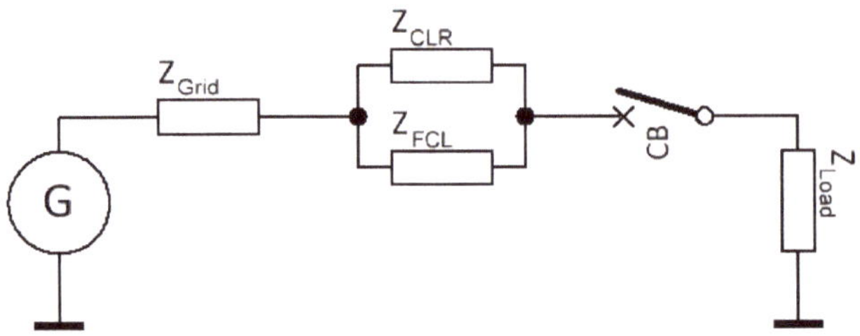

Figure 4.5.1: Exemplary representation of the relevant main impedances to illustrate how the current limiter (Z_{FCL}) and current limiting reactor (Z_{CLR}) are arranged in relation to each other

In normal operation, in which the fault current limiter is in superconducting mode, nearly the entire current flows through the fault current limiter, so that the current limiting reactor is designed for a negligibly small nominal current of only a few hundred amperes. If a fault occurs and the fault current limiter quenches, i.e. loses the superconducting state and gets its high electrical resistance, the prospective short-circuit current is limited to its design value. During limitation the fault current limiter has a higher resistance than the current limiting reactor. Thus, the fault current commutates to the parallel path of the current limiting reactor until a fast circuit breaker (CB) clears the failure. For the current limiting reactors of case A and B, this operating scenario results in the requirements given in Table 4.5.1.

Table 4.5.1: General technical design requirements for the CLRs

Property	Case A	Case B
Nominal Voltage, U_n	380 kV	
Highest Voltage for Equipment	440 kV	
Nominal Current, $I_{n,CLR}$	Few hundreds of Ampere	
Max. limited peak current with SFCL, I_k''	28 kA	42 kA
Frequency f	50 Hz	
Fault duration t_d	0.25 s	
Rated Inductance L_{CLR}	72.9 mH	39.5 mH
Unified specific creepage distance	34.7 mm/kV$_{(ph-gr)}$	
Wind load	34 m/s	
Seismic demands	0.5 g	
Altitude	1000 m.a.s.l.	
Max. ambient temperature	40 °C	
Max. daily average temperature	35 °C	
Location	Outdoor	
Quantity	3	
Applicable standard	IEC60076-6	

The high-voltage requirements for the fault current limiter, specified in Table 2.2.2, are also applied to the current limiting reactors.

Based on the technical requirements according to Table 4.5.1, the resulting characteristics and dimensions of the current limiting reactors are shown in Table 4.5.2 as well as in Fig. 4.5.2 and Fig. 4.5.3.

Table 4.5.2: Technical properties of the current limiting reactors

Property	Case A	Case B
Design technology	Air-core, dry-type reactors	
Number of coils per phase	1	
Number of coils per reactor unit	1	
Number of phases per reactor unit	1	
Winding conductor material	Aluminium	
Mounting arrangement	Side-by-side	
Type of insulators	Porcelain	
Protection class	IP 00	
Pollution level acc. to IEC60815-1	Medium	
Minimum insulator creepage distance to ground	8815 mm	
Total mass (incl. insulators)	≈ 10700 kg	≈ 12600 kg
Rated reactance X_{CLR}	22.092 Ω	12.409 Ω
Max. thermal short circuit current (RMS) according to IEC 60076-6 for 0.25 s	11.04 kA	16.47 kA
Coil distance D_{cc} (center-to-center)	6260 mm	6330 mm
Height of magnetic clearance HMC1	0	0
Height of magnetic clearance HMC2	0	0
Radius of magnetic clearance RMC1	1600	0
Radius of magnetic clearance RMC2	1910	1770

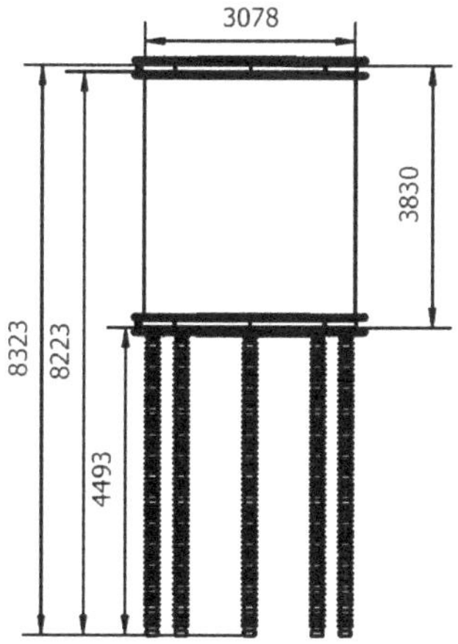

Figure 4.5.2: Dimension of the current limiting reactor for Case A

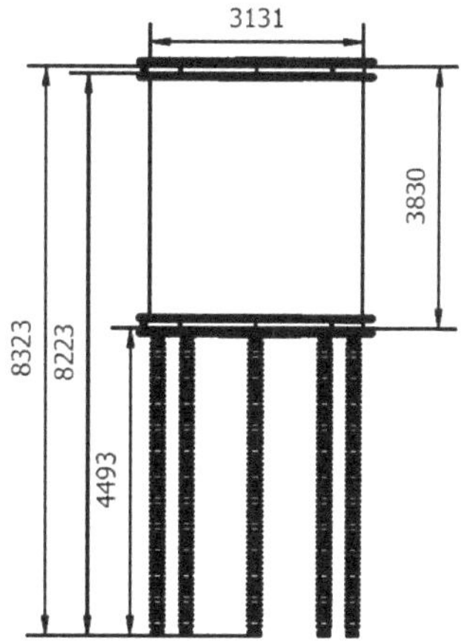

Figure 4.5.3: Dimension of the current limiting reactor for Case B

When installing the current limiting reactors in the substation, care must be taken to ensure that a minimum distance D_{cc} is maintained between the individual current limiting reactors (see Fig. 4.5.4). This is around 6.4 m for both applications, as shown in Table 4.5.2.

Finally, the distances to metallic components in the vicinity of current limiting reactors must be designed. As shown in Fig. 4.5.5, HMCx (height of magnetic clearance contour) defines the upper and lower area outside the coil body and RMCx (radius of magnetic clearance contour) the radial distance from the centre of the coil body. There are two areas to consider. There must be no metal parts in the area of HMC1 and RMC1. For larger metal structures, such as rebar, earth wires and fences, HMC2 and RMC2 clearances should be observed. These values can be found in Table 4.5.2.

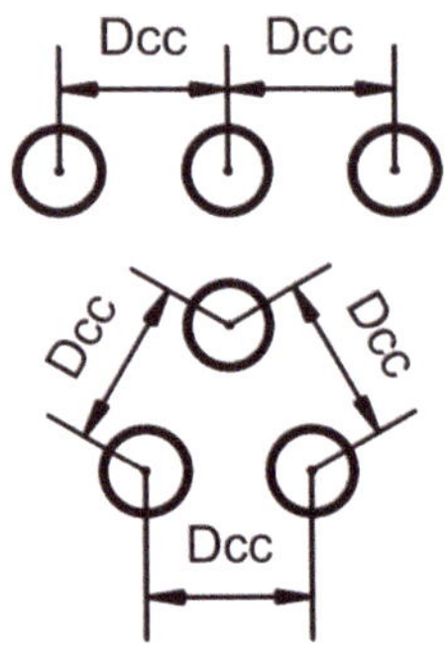

Figure 4.5.4: Common arrangements (line and triangle) of CLRs and the minimum distance to be maintained

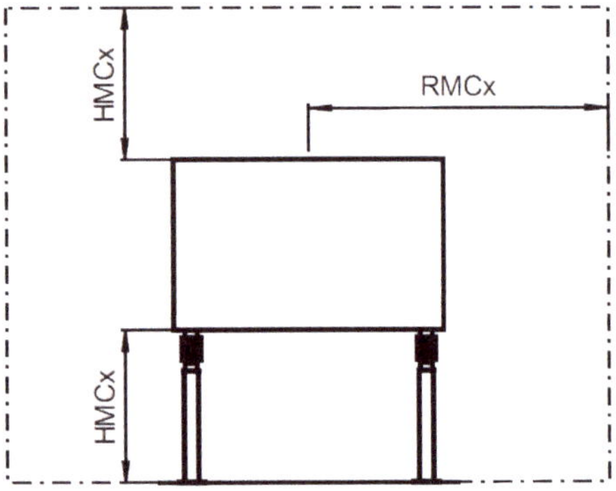

Figure 4.5.5: Magnetic clearance contour per phase of CLR

The reason for the relatively small distances is that they are based on the nominal current. Since this is negligibly low in this particular case, there is no significant minimum distance for metallic components.

4.6 Loss Estimation

The total losses of a resistive current limiter are mainly determined by the heat input at low temperatures. In the case of the superconducting current limiter, this heat input at low temperatures P_{77K} is made up of several components. Although superconductors do not have a DC resistance, so-called AC losses P_{AC} occur under alternating field or alternating current stress, which is caused by the movement of the flux vortices in the superconductor [Wilson1987]. Although these are small compared to the losses in normal conductors, they can play an important role in the dimensioning of the cooling system. Another large heat input is due to the heat conduction and current heat losses of the current leads P_{CL}. The heat input by the cryostat P_{Cryo} with significant loss components due to thermal radiation and the mechanical support structures must also be taken into account. The contacts between copper and superconductor must be made with very low resistance, but since a large number occur, the total losses due to contacting in the current limiter $P_{Cu\text{-}HTS}$ are also taken into account here. Other losses P_{Other} can occur through measuring leads of sensors. Thus, the heat input to low temperatures P_{77K} in the current limiter adds up to

$$P_{77\mathrm{K}} = P_{\mathrm{AC}} + P_{\mathrm{CL}} + P_{\mathrm{Cryo}} + P_{\mathrm{Cu-HTS}} + P_{\mathrm{others}} \qquad (4.6.1)$$

The dependencies of the individual types of losses are described in the following and the losses as a function of the current are calculated for the various design cases.

4.6.1 AC Losses in the HTS Tapes

AC losses in a bifilar array of HTS tapes depend on various parameters and increase:

- with increasing distance between forward and return conductors d_{tt},
- with increasing width of the HTS tape b and
- with increasing current I through the superconductor.

Table 4.6.1 shows the very strong dependence of the losses on the distance between the forward and return conductors d_{tt} and the current I. In principle, a larger amount of superconductor in the parallel circuit could also reduce the AC losses. Furthermore, according to the equation of Norris [Norris1970], the transport current losses of an HTS tape depend quadratically on the critical current I_c and quite complexly on the ratio of the actual current to the critical current. Since the critical current ideally increases linearly with the width of the HTS tape, losses consequently also increase with the square of the width. In the bifilar arrangement, however, the ratios are more complex, but even in this arrangement the losses increase disproportionately with the width of the tape.

Table 4.6.1: Specific AC loss of a 12 mm wide HTS tape with a critical current of 600 and 700 A (all values in W/km)

d_{tt}	1 mm	2 mm	3 mm	4 mm	6 mm
1 · I_c	43.6 / 59.3	171 / 233.1	375.9 / 511.6	627.4 / 853.9	1134 / 1544
0.9 · I_c	19.8 / 26.9	71 / 96.6	145.1 / 197.5	233.6 / 318	424.3 / 577.5
0.8 · I_c	10.7 / 14.6	37.3 / 50.8	75.1 / 102.2	119.5 / 162.6	214.4 / 291.8
0.7 · I_c	5.7 / 7.8	19.7 / 26.8	39.1 / 53.2	61.6 / 83.9	109.7 / 194.4
0.6 · I_c	2.9 / 4.0	9.8 / 13.4	19.3 / 26.3	30.4 / 41.4	53.8 / 73.2
0.5 · I_c	1.4 / 1.9	4.5 / 6.1	8.8 / 12	13.7 / 18.7	24.2 / 32.9
0.4 · I_c	0.59 / 0.76	1.8 / 2.4	3.5 / 4.7	5.4 / 7.3	9.4 / 12.8
0.3 · I_c	0.51 / 0.69	0.56 / 0.76	1.1 / 1.5	1.7 / 2.3	2.9 / 3.9

The losses decrease strongly with smaller distance between the forward and return conductors. To minimize losses, one would try to place the conductors close together. However, this makes cooling with liquid nitrogen more difficult and increases the electric field between the conductors. Thus, a value for the distance between the forward and return conductors must be found here that meets all requirements.

Although it would be possible to reduce the AC losses even further by using a 4 mm wide HTS tape, this would increase the number of coils by a factor of three and would involve an additional design effort that would not make sense. For resistive superconducting current limiters, therefore, HTS tapes with a width of 10-12 mm are usually used.

The AC losses in Table 4.6.1 are calculated using the approximate equation for bifilar coils given in [Clem2008] and the setup in Fig. 4.6.1.

$$P = f\mu_0 I_c^2 \left[\frac{2}{\pi a^2} \int_c^a (a-x)\tanh^{-1}\sqrt{\frac{\sinh^2\frac{\pi x}{D} - \sinh^2\frac{\pi c}{D}}{\sinh^2\frac{\pi a}{D} - \sinh^2\frac{\pi c}{D}}}\,dx + \frac{d}{12a}\left\{ 1 - \frac{c}{a} + \frac{8}{\pi^3 a}\int_0^c \left[\tan^{-1}\sqrt{\frac{\sinh^2\frac{\pi a}{D} - \sinh^2\frac{\pi c}{D}}{\sinh^2\frac{\pi c}{D} - \sinh^2\frac{\pi x}{D}}} \right] dx \right\} \right]$$

In this equation *P* are the losses per m, *f* = 50 Hz the frequency, $D = d_{tt}$ the distance between adjacent tapes and *d* is the thickness of the tape, or, in our case, of a twin. The actual flowing current *I* is hidden in a nontrivial way in the parameter *c* [Clem2008].

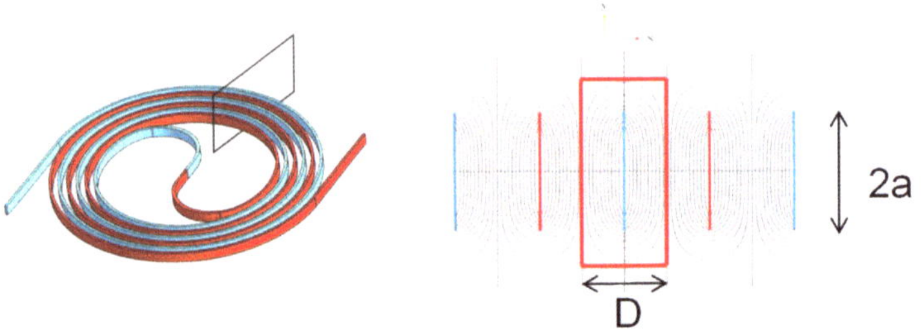

Figure 4.6.1: Bifilar coil (left) with main dimensions (right)

Table 4.6.2 shows the total AC losses of the HTS tapes used for different currents and distances between the forward and return conductors. As the distance increases, the perpendicular field component on a conductor increases and so do the losses. Since the losses depend very much on the current, one measure to reduce the losses could be to connect more HTS tapes in parallel. However, since this changes the current limitation, leads to additional costs for the superconductor and to larger cryostat dimensions, an attempt is made to do without this measure.

Table 4.6.2: AC loss per phase for d_{tt} = 2 mm, I_c = 700 A

	Case A 78.4 km/phase	**Case B 69.7 km/phase**
$1 \cdot I_c$	18.275 kW	16.247 kW
$0.9 \cdot I_c$	7.573 kW	6.733 kW
$0.8 \cdot I_c$	3.982 kW	3.540 kW
$0.7 \cdot I_c$	2.100 kW	1.868 kW
$0.6 \cdot I_c$	1.050 kW	0.933 kW
$0.5 \cdot I_c$	0.478 kW	0.590 kW
$0.4 \cdot I_c$	0.188 kW	0.167 kW
$0.3 \cdot I_c$	0.059 kW	0.053 kW
$0 \cdot I_c$	0 kW	0 kW

Since the HTS tapes are not operated at the critical current density the AC losses shown in the previous Table need to be adjusted to the actual ratio of the peak current to the critical current.

This is done in Table 4.6.3, which shows the total AC loss per phase for the selected module arrangement shown in chapter 4.3.2. It is typical for superconductors that the AC losses increase strongly near the critical current. In general, they are related to the current with the power of three to five. A maximum AC loss of 5.1 kW for case A and 4.5 kW for case B was calculated and taken into account for the design of the cooling system.

Table 4.6.3: AC loss per phase for d_{tt} = 2 mm, I_c = 700 A (e.g. 7071 A / 8400 A = 0.84178571)

Current	Factor I/I_c	Specific loss	Total loss Case A 78.4 km/phase	Total loss Case B 69.7 km/phase
$1 \cdot I_r$	$1 \cdot 0.841 = 0.841$	65.9 W/km	5.166 kW	4.593 kW
$0.9 \cdot I_r$	$0.9 \cdot 0.841 = 0.7569$	38.6 W/km	3.026 kW	2.690 kW
$0.8 \cdot I_r$	$0.8 \cdot 0.841 = 0{,}6728$	22.3 W/km	1.748 kW	1.554 kW
$0.7 \cdot I_r$	$0.7 \cdot 0.841 = 0.5887$	12.3 W/km	0.964 kW	0.857 kW
$0.6 \cdot I_r$	$0.6 \cdot 0.841 = 0.5046$	6.3 W/km	0.493 kW	0.439 kW
$0.5 \cdot I_r$	$0.5 \cdot 0.841 = 0.4205$	3.0 W/km	0.235 kW	0.210 kW
$0.4 \cdot I_r$	$0.4 \cdot 0.841 = 0.3364$	1.2 W/km	0.094 kW	0.084 kW
$0.3 \cdot I_r$	$0.3 \cdot 0.841 = 0.2523$	0.675 W/km	0.053 kW	0.047 kW
$0 \cdot I_r$	0	0 W/km	0 kW	0 kW

Since Clem's formula is not foreseen for a bifilar coil arrangement of HTS tapes, FEM calculations are performed and compared with Clem's formula. A twin tape in bifilar configuration was compared with Clem's formula for a (thicker) single tape in bifilar configuration. Table 4.6.4 lists the cyclic AC losses (in J/m) for different current ratios. The separation between the centers of the bifilar turns is 2 mm. The agreement is good, especially in the current range 0.6-0.8 of the critical current I_c. Clem's formula diverges and gives higher values at $I=I_c$, so it is not surprising that the agreement worsens e.g. at 0.9 I_c. Since the FEM results show lower AC losses than the losses calculated with Clem's formula, it can be concluded that the previous SFCL AC loss estimations derived from Clem's formula are on the safe side.

Table 4.6.4: AC loss comparison for 12 mm, 600 A, 50 Hz, 3 µm, n-value 25 (all values in J/m)

I/I_c	FEM twins	Clem
0.9	4.71e-3	5.73e-3
0.8	2.77e-3	2.92e-3
0.7	1.58e-3	1.52e-3
0.6	8.50e-4	7.95e-4

4.6.2 Heat Input through the Current Leads

The current lead connects the superconductor at low temperatures to the terminal of the current limiter at ambient temperature. As Fig. 4.6.2 shows, there are two main loss fractions, for a given current I and length l, that behave in opposite directions to the diameter of the current lead.

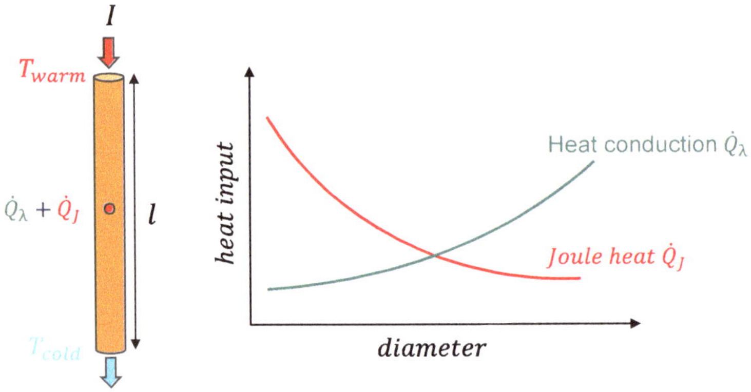

Figure 4.6.2: General scheme of a conduction cooled current lead (left) and general heat input as a function of the current lead diameter (right)

The heat input due to conduction $\dot{Q}_\lambda$ increases with increasing diameter D of the current lead, while the heat input due to Joule's heat loss $\dot{Q}_J$ decreases.

$$\dot{Q}_\lambda = \frac{\lambda(T)\,\pi\left(\frac{D}{2}\right)^2}{l}\,\Delta T \qquad (4.6.3)$$

$$\dot{Q}_J = \frac{\rho(T)\,l}{\pi\left(\frac{D}{2}\right)^2}\,I^2 \qquad (4.6.4)$$

Table 4.6.5: Total heat input for optimized conduction cooled current lead and different rated and load currents (42.45 W/m, three phases with 6 current leads in total)

Load	**5000 A**
$1 \cdot I_r$	1274 W
$0.9 \cdot I_r$	1162 W
$0.8 \cdot I_r$	1071 W
$0.7 \cdot I_r$	997 W
$0.6 \cdot I_r$	937 W
$0.5 \cdot I_r$	889 W
$0.4 \cdot I_r$	852 W
$0.3 \cdot I_r$	823 W
$0.2 \cdot I_r$	800 W
$0.1 \cdot I_r$	788 W
$0 \cdot I_r$	784 W

Using the well known values of temperature-dependent thermal conductivity $\lambda(T)$, resistivity $\rho(T)$ (Annex 12.2), and temperature difference ΔT between the hot and cold ends of the copper current lead, the heat input can be calculated [Buyanov1975]. The heat input can be optimized only for a given current value. According to the Wiedemann-Frantz law [Wiedemann-

Frantz1853], an optimal length to cross section ratio for a given current value results in a minimized heat input.

For copper with a residual resistance ratio (RRR) of 50, a purely conduction-cooled current lead results in a minimum heat input of 42.45 W per kA from ambient temperature to 77 K. As an example, assuming a length of one meter, the optimum diameter is D = 19.09 mm. A conduction-cooled current lead represents the simplest case in terms of design. If in addition the enthalpy of the coolant is used to cool the current lead, the heat input can be reduced to 23.4 W per kA and thus roughly halved. If longer copper connections are required within the liquid nitrogen, their loss ratio can be reduced by connecting them in parallel with HTS tapes.

4.6.3 Heat Input through the Cryostat

A double-walled cryostat with multilayer insulation and high vacuum usually has a heat input from ambient temperature to the temperature of the liquid nitrogen of about 1 W per square meter of surface [VDI2013]. Due to the high vacuum, the residual gas heat conduction is negligible, so that the resulting heat input is caused by thermal radiation. This also explains the use and benefit of reflective multilayer insulation. Further heat input by thermal conduction is caused by the mechanical support between the outer and inner walls and by the flanges. The flanges are required for insertion of the limiter module. However, this input can be kept low by a special mechanical design. Only a precise design of the cryostat can determine these values exactly. Since these heat inputs increase with the size of the cryostat, a heat input of 2 W/m² is assumed for simplification in this study. Thus, the influence of the mechanical supports and the flanges is also estimated.

Table 4.6.6: Heat input to cryostat from 300 K to 77 K (2 W/m²)

r_i	L = 10 m	L = 12 m	L = 14 m
1.5 m	218 W	254 W	292 W
1.6 m	233 W	273 W	314 W
1.7 m	250 W	293 W	335 W

4.6.4 Heat Input through Contacting and Others

The contact of the HTS tape with copper can be made with very low impedance by soldering. Usually, values in the nano-ohm range can be achieved with standard conductors from various manufacturers. However, these values can vary greatly depending on the material and design. With a conservative assumption of 0.5 µΩ per superconductor to copper contact, this results in a power dissipation of 2.45 mW at 700 A per superconductor tape. A module with 6 conductors in parallel then has a maximum of 12 contacts and thus a power dissipation at a rated current of 29.4 mW for 700 A. Since no variant has more than 100 modules installed, the additional loss contribution due to contacting would be less than 2 W and thus within the uncertainty range of the larger loss contributions.

Further heat contributions can arise from the attachment of sensors. These are assumed to be comparatively small. In addition, the high voltage bushings with their internal aluminum layers cause losses. To simplify matters, the heat input due to contacting and other losses are thus assumed to be 20 W in this study. This value includes the copper resistance between the modules. As can be seen in the following summary, this heat input is relatively low compared to the previous ones and thus not of high importance.

4.6.5 Heat Input Summary

Table 4.6.7 and Table 4.6.8 summarize the total heat input for cases A and B respectively. It can be seen that for the maximum current, the heat input is dominated by the AC loss and that for average currents, the contributions of the current lead and the cryostat get more important. These values are the basis for the design of the cooling system, which follows, in the next sub-chapter.

Table 4.6.7: Total heat input and cooling power (three phase) from environmental temperature 300 K to 77 K for Case A

Load	AC loss	Current Lead	Cryostat	Others	Total
$1 \cdot I_r$	$3 \cdot 5.166$ kW	1274 W	~ 900 W	20 W	17.692 kW
$0.9 \cdot I_r$	$3 \cdot 3.026$ kW	1162 W	~ 900 W	20 W	11.160 kW
$0.8 \cdot I_r$	$3 \cdot 1.748$ kW	1071 W	~ 900 W	20 W	7.235 kW
$0.7 \cdot I_r$	$3 \cdot 0.964$ kW	997 W	~ 900 W	20 W	4.809 kW
$0.6 \cdot I_r$	$3 \cdot 0.493$ kW	937 W	~ 900 W	20 W	3.336 kW
$0.5 \cdot I_r$	$3 \cdot 0.235$ kW	889 W	~ 900 W	20 W	2.514 kW
$0.4 \cdot I_r$	$3 \cdot 0.094$ kW	852 W	~ 900 W	20 W	2.054 kW
$0.3 \cdot I_r$	$3 \cdot 0.053$ kW	823 W	~ 900 W	20 W	1.902 kW
$0 \cdot I_r$	0 W	800 W	~ 900 W	20 W	1.720 kW

Table 4.6.8: Total heat input and cooling power (three phase) from environmental temperature 300 K to 77 K for Case B

Load	AC loss	Current Lead	Cryostat	Others	Total
$1 \cdot I_r$	$3 \cdot 4.593$ kW	1274 W	~ 900 W	20 W	15.973 kW
$0.9 \cdot I_r$	$3 \cdot 2.690$ kW	1162 W	~ 900 W	20 W	10.152 kW
$0.8 \cdot I_r$	$3 \cdot 1.554$ kW	1071 W	~ 900 W	20 W	6.653 kW
$0.7 \cdot I_r$	$3 \cdot 0.857$ kW	997 W	~ 900 W	20 W	4.488 kW
$0.6 \cdot I_r$	$3 \cdot 0.439$ kW	937 W	~ 900 W	20 W	3.174 kW
$0.5 \cdot I_r$	$3 \cdot 0.210$ kW	889 W	~ 900 W	20 W	2.439 kW
$0.4 \cdot I_r$	$3 \cdot 0.084$ kW	852 W	~ 900 W	20 W	2.024 kW
$0.3 \cdot I_r$	$3 \cdot 0.047$ kW	823 W	~ 900 W	20 W	1.884 kW
$0 \cdot I_r$	0 W	800 W	~ 900 W	20 W	1.720 kW

4.7 Cryogenic Design

4.7.1 Introduction

Reliable operation of superconducting current limiters requires continuous cooling with liquid nitrogen at an operating temperature of about 77 K. A lower temperature of, for example, 68 K would increase the critical current density of the superconductor by more than a factor of two and correspondingly reduce the superconductor quantity and cost. However, our own investigations within the EU project FASTGRID [Tixador2019] revealed a frequently occurring conductor degradation, which is due to the significantly increased energy density in the superconductor at the moment of transition into the normal line. Since this problem has not yet been clarified with sufficient certainty, the operating temperature of the superconductor is assumed to be 77 K in this design.

A basic distinction is made between open and closed cooling methods for superconducting current limiters, similar to superconducting cables. In an open system, liquid nitrogen must be continuously supplied from a storage tank. This must then be refilled at regular intervals. This has the lowest investment costs compared with the closed cooling cycle, and a very high reliability result, since the redundancy can be realized by a multiple arrangement of simple vacuum pumps. In the AmpaCity project in Essen, Germany, this method was used to reliably operate a 10 kV, 40 MVA, 1 km long cable with a current limiter for more than 7 years.

At 77 K one kW of heat evaporates approximately 20 kg of liquid nitrogen per hour or 480 kg per day at normal pressure conditions. In an open cooling system, it depends on the size of the liquid nitrogen reservoir how often a refill needs to be done. With the total heat input given in the previous sub-chapter a refill within a few days or one week is needed.

In a closed system, a cryocooler is used to recool the nitrogen. In comparison with the open system no regular refill of the liquid nitrogen reservoir is needed. The cryocoolers under consideration differ mainly in their cooling principle, cooling capacity, and efficiency [Radebaugh2004]. Table 4.7.1 shows a summary of typical cryocoolers with the most important data.

Table 4.7.1: Different cryocooler with main data

Type	**AL 600**	**PTC 1000**	**SPC-1**	**Brayton Neo**
Manufacturer	Cryomech	AFCRYO	Stirling Cryogenics	Mayekawa
Principle	GM	Pulse tube	Stirling	Turbo Brayton
Cooling power @ 77 K	600 W	1450 W	1200 W	5000 W
Efficiency	5.5 %	4.8 %	11 %	8 %
Dimensions mm^3	610x610x790	1730x1180x1940	860x620x1320	2200x3600x2200
Maintenance interval	10000 h	4 a	6000 h	> 15000 h

At high voltages of 380 kV the liquid nitrogen has to be kept under pressure to improve the breakdown voltages. At higher pressures than normal pressure, liquid nitrogen has an equilibrium temperature above 77 K. The pressure must be controlled and the necessary cooling capacity must be supplied via a cryocooler with a heat exchanger. To keep the temperature difference in the cryostat low, a flow of liquid nitrogen is also required. Liquid nitrogen pumps are required for this purpose.

Presently, various open and closed cooling methods are basically possible for use in superconducting current limiters and have already been implemented. The longest operating experience is with the open system at 1 bar [Stemmle2015]. The recently commissioned 220 kV current limiter [Moyzikh2021] uses a closed system with an operating pressure of 5 bar.

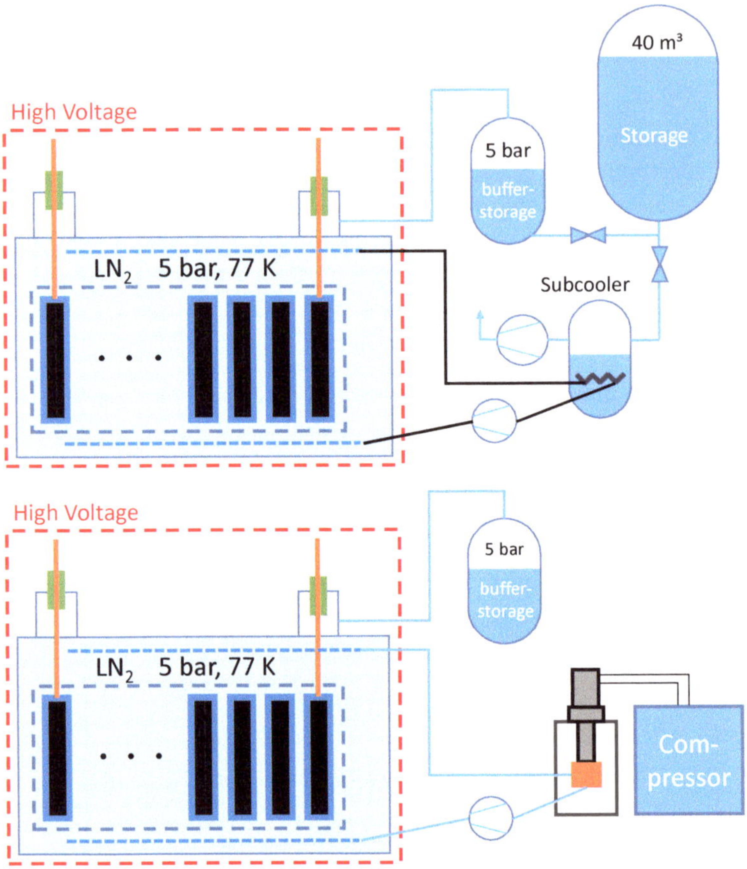

Figure 4.7.1: Sketch of open (top) and closed (bottom) cooling methods for fault current limiters with pressurized liquid nitrogen

For open cooling at a pressure of 5 bar (Fig. 4.7.1 top), additional containers, a heat exchanger and additional vacuum or liquid nitrogen pumps are required. The electrical power of the vacuum pumps $P_{el,vac}$ and the electrical power of the liquid nitrogen pumps $P_{el,LN2}$ are added to the consumption of the liquid nitrogen during operation.

$$P_{el} = P_{el,vac} + P_{el,LN2} \qquad (4.6.1)$$

In addition to the previously calculated heat input P_{77K}, an additional heat input by the liquid nitrogen pumps $P_{LN2pump}$ must also be taken into account. However, this is so small compared to the other inputs that it is neglected in this study. The pumps require regular maintenance but can be serviced without interrupting operation as shown in the AmpaCity cable operation. Thus, a similar high reliability as in the open system at 1 bar can be expected.

In the case of closed cooling with 5 bar (Fig. 4.7.1 bottom), a corresponding cryocooler can be integrated in the primary circuit as shown. The required electrical power is then given by:

$$P_{el} = \frac{1}{\eta} P_{77K} + P_{el,LN2} \qquad (4.6.2)$$

In this case, the temperature of the liquid nitrogen can be controlled very easily by the cryocooler. The operating costs are calculated by the purchase price for the electrical energy.

4.7.2 Description of Cryocooler Open Cooling System

The facility specified below is designed to maintain a cryogenic coolant loop for the cold energy supply of up to 16 kW (average: 4.5 kW) for high temperature superconductor devices. It provides a flow rate of up to 1.5 kg/s of sub-cooled liquid nitrogen at an outlet temperature of 71 ± 0.25 K. This can be adapted to 77 K as well.

The plant utilizes liquid nitrogen as both the source of cryogenic energy and the circulated coolant fluid. The liquid is withdrawn from the bottom interface of a vacuum-insulated cryogenic storage vessel (to be offered by local provider) that is also equipped with a high capacity pressure build-up system, a redundant pressure regulation system with a blow-out silencer and a set of high flow Pressure Safety Valves (PSV). Within the coolant loop, the liquid nitrogen tank also serves as a buffer vessel compensating temperature and pressure-related volumetric variations.

The cooling system provides three individual coolant loop flows of 0.5 kg/s of sub-cooled liquid nitrogen to three SFCL recipient vessels. The subcooler contains three individual heat exchangers that are submerged in a liquid nitrogen boiling bath at a temperature of approx. 68 K (sub-cooler units). The low temperature is achieved by adjusting the pressure above the nitrogen bath to approx. 280 mbar by means of a set of three high capacity dry running vacuum pumps.

High efficiency cryogenic reciprocating pumps are implemented for the continuous circulation of the coolant through the heat exchangers and fault current limiters. A redundant cryogenic filter system (50 μm mesh) is located in each loop downstream the sub-cooler with valves and instrumentation for an automated switch-over in case of an excessive pressure drop. It protects fault current limiter devices and process equipment against particle pollution.

In order to tolerate any single component failure of items with a service life of 10 years or less without service interruption the system is designed with redundant process equipment and corresponding control and monitoring functions. All components with a limited service life can be serviced or repaired while the coolant loop is in operation. (Process equipment like heat exchangers - including the sub-cooler - as well as transfer piping and valve bodies are considered to have a service life exceeding 10 years.)

As far as feasible, without limitation of the serviceability, the cryogenic equipment and the related transfer system including valves, filters, instruments, pressure devices, pumps and interfaces are vacuum-insulated in order to minimize the external heat ingress and the related nitrogen consumption. Line sections and integrated valves which are only used temporarily for chill-down purposes are not provided with a vacuum jacket.

Beside the equipment monitoring features and the pressure control of the vacuum system, the Programmable Logic Controller manages the tank pressure control and the steady level control of the coolant bath. Furthermore, it provides for the human-machine interfacing facilitating routine inspections, maintenance and the required commissioning and failure handling functions. As an option, a fully fail-safe Programmable Logic Controller is offered.

The plant is supplied in a container enclosure suitable for outdoor set-up (mild steel with corrosion protection) with noise insulation, ventilation, lighting and internal atmosphere conditioning (including oxygen concentration monitoring). The system is factory-tested and pre-installed ready for the interface connection on site.

Basic functionalities are shown in the simplified diagram in Fig. 4.7.2.

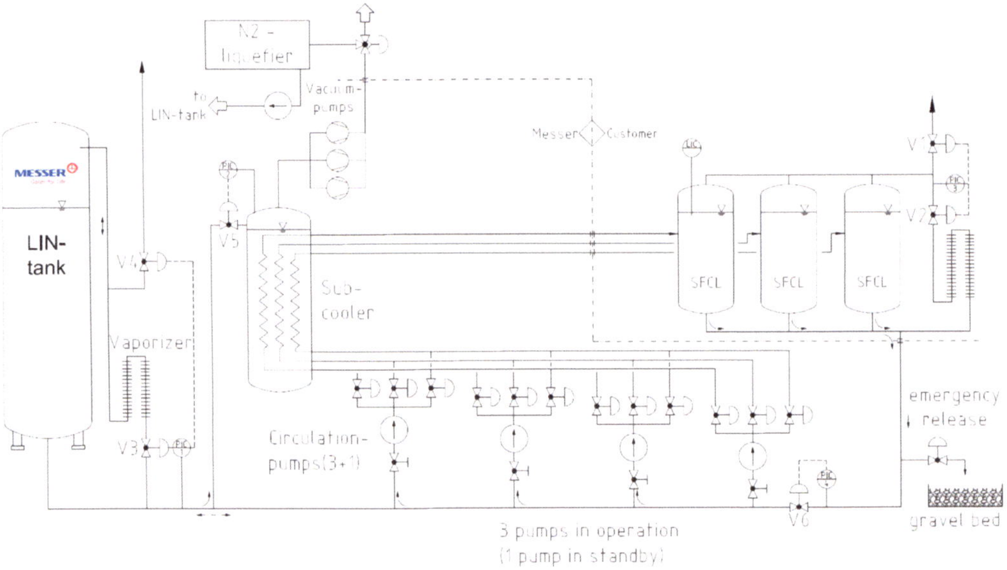

Figure 4.7.2: Simplified piping and instrumentation diagram of the open cooling circuit for a three-phase superconducting fault current limiter (Source: F. Herzog, Messer)

To reduce liquid nitrogen consumption gaseous nitrogen leaving the system can optionally be liquefied and pumped back into the storage vessel by an external N_2-liquefier. If such a system is installed the described cooling system serves as redundancy in case of malfunction or maintenance of the liquefier.

The cooling system is mounted in a steel container (app. 3 x 3 x 8 m) including noise insulation, ventilation and lighting, internal atmosphere conditioning including oxygen concentration monitoring and various service openings for easy maintenance.

Usually the storage vessel and delivery of liquid nitrogen will be offered by a provider of liquid nitrogen in a separate agreement. The liquid nitrogen consumption calculated includes the heat ingress of the storage vessel.

Table 4.7.2: Main data for the cooling system

Process liquid / coolant	Liquid nitrogen
Tank operating pressure	5 bar up to 8 bar
Service pressure HTSC inlet	max. 1 to 2 bar above tank pressure
Generated cold max. (Ø)	18 (6) kW @ 68 K (reserve for int. losses)
Delivered cold max (Ø)	16 (4.5) kW @ 71 K (available for SFCLs)
SFCL inlet temperature	< 71 K
SFCL return temperature	App. 77 K
Coolant flow	3 x 0.5 kg/s
Average liquid nitrogen consumption	app. 120 kg/h (@ 4.5 kW usable cold)
Operating mode	Continuous
Power supply	3 x 400 V - 50 Hz, appr. 100 A
Instrument pressure supply (integr.)	6 bar (g)

In an open cooling system as shown above, only the vacuum and liquid nitrogen pumps require regular maintenance only. Since several pumps are connected in parallel, the operation of the SFCL is not affected during the maintenance or outage of the pumps. After the setup and filling of the cryostat and the liquid nitrogen reservoir, the first commissioning of the system can be done within less than a week. The main work for the first commissioning includes checking the leak tightness, making vacuum, ventilating with gaseous nitrogen and gradually filling the system with liquid nitrogen. Finally, a stable operation of the system for a few days is recommended before a grid connection of the SFCL.

4.8 System Layout

Based on the results of the previous chapters, the system layout of the 380 kV SFCL is summarized in this sub-chapter.

4.8.1 HTS Module

The main parameters and a design drawing of the HTS module are shown in Table 4.8.1 and Fig. 4.8.1. One HTS module consists of 6 twin tapes in parallel, each with a length of 76.6 m for case A and 62.4 m for case B.

Table 4.8.1: Main parameters of the HTS modules and current limiting element

	Case A	Case B
Number of HTS modules	85	93
Number of HTS module connected in series	85	93
Outer diameter of HTS module without corona ring	1100 mm	1000 mm
Outer diameter of HTS module with corona ring	1300 mm	1300 mm
Total length of limiting element	6.38 m	7.03 m
Number of HTS tapes in parallel	12 (6 twin tapes)	12 (6 twin tapes)
Module voltage at limitation*	18.5 kV	16.9 kV
Single piece length of HTS tape within one HTS module	76.6 m	62.4 m
Total HTS length within one HTS module	12 x 76.6 m	12 x 62.4 m

* assuming homogenous voltage distribution

The HTS tape is held in position with a GFK disk at both sides. This arrangement allows a good cooling with liquid nitrogen, whilst giving a good mechanical stability and compensates for the elongation due to a higher temperature at current limitation. In addition, the contacts to the copper connections can be distributed around the HTS module.

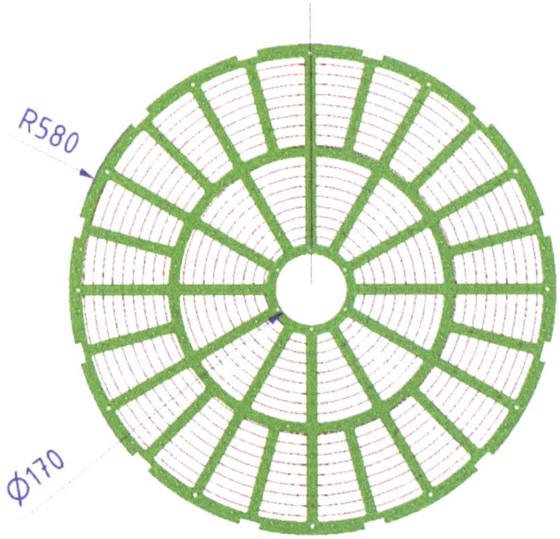

Figure 4.8.1: Drawing of one HTS module without corona ring for case A. Only one HTS twin tape is shown.

4.8.2 Single Phase SFCL

A sketch of one phase of the 380 kV SFCL is shown in Fig. 4.8.2 and the main parameters are given in Table 4.8.2. Two options for the cryostat were investigated. Firstly, the gas reservoir is placed in a separate vessel besides the cryostat. This would be the most compact design but during the limitation, the pressure increase in the cryostat is higher. A calculation of the pressure increase during limitation is very complex and not within the scope of this study. Secondly, a gas reservoir is located within the cryostat. We assumed 20 % of the total volume as gas reservoir. In addition, it is possible to position the gas reservoir in a dome on top of the cryostat but this version would require a more complex design of the cryostat and is therefore not considered in this study.

Table 4.8.2: Main data of a single phase of a 380 kV SFCL

	No gas reservoir	**With 20 % gas reservoir**
Outer dimensions (L/W/H)	13664 mm / 3640 mm / 10255 mm	13664 mm / 4140 mm / 10302 mm
Height without bushing	4940 mm	5038 mm
Total weight without / with liquid nitrogen	48 tons / 142 tons	53 tons / 165 tons
Amount of liquid nitrogen	94 tons	112 tons

A single phase of a 380 kV SFCL is transported without bushing and liquid nitrogen filling. With a combination of ship and road transport any location within Germany should be accessible. Companies for the ship and road transport were not contacted within the scope of this study.

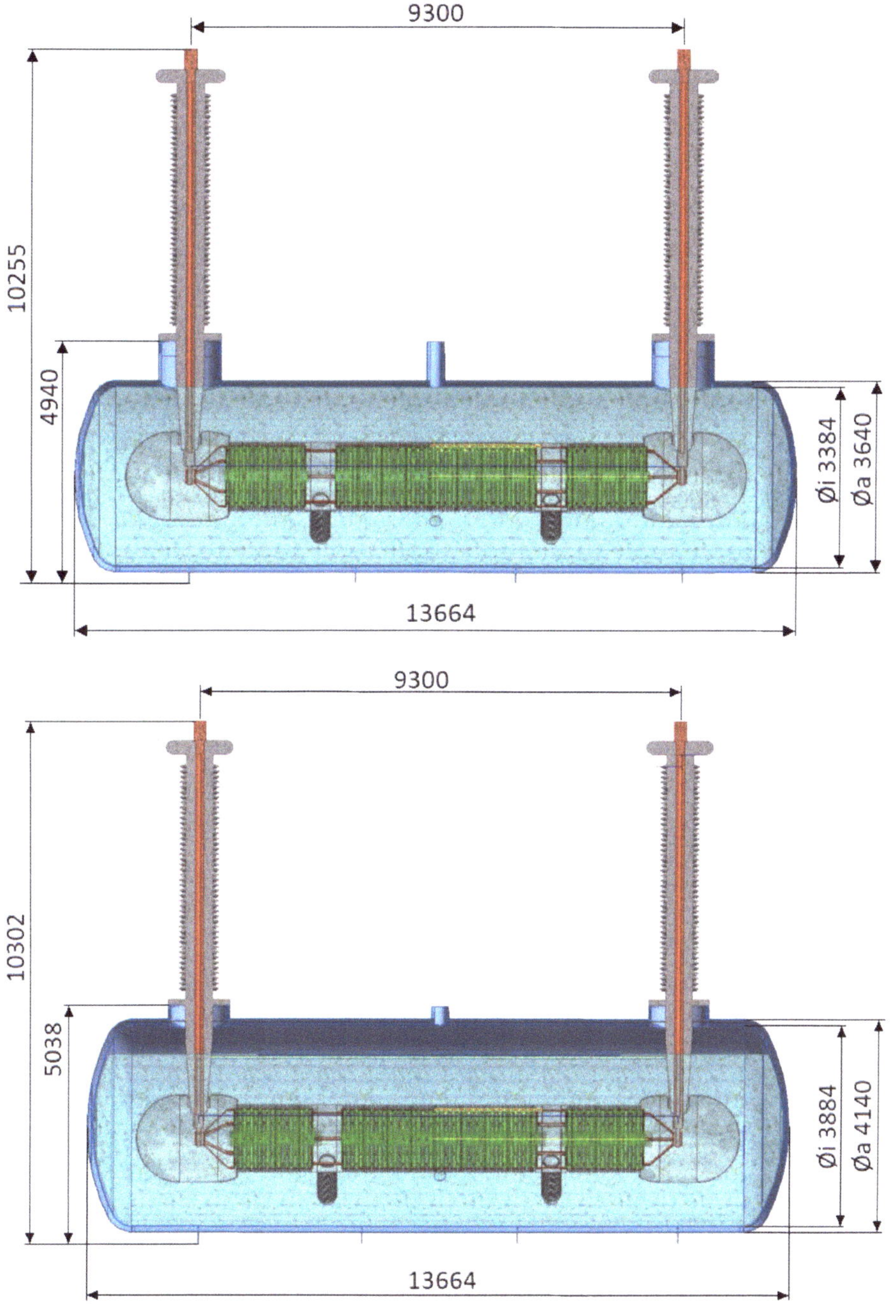

Figure 4.8.2: Sketch of one phase of a 380 kV SFCL for case A and case B (top: version without gas reservoir in cryostat, bottom: version with 20 % gas reservoir in cryostat)

4.8.3 General Layout

The total 380 kV SFCL system consists of three SFCLs plus three current limiting reactors in parallel connection plus a standard 20 ft container (6.10 m long x 2.44 m wide x 2.59 m high) for an open cooling system or two standard containers for a closed cooling system. Possible arrangements of the three SFCLs and the three current limiting reactors are shown in Fig. 4.8.3 and Fig. 4.8.4. The detailed setup of the SFCL and the current limiting reactor including their connections depends on the situation within the 380 kV station and is not within the scope of this study. It has been checked that the proposed size fits into such a station.

The location of the standard container of the cooling system can be chosen arbitrarily because the distance from the cooling to the SFCL can be up to 100 m and even more. A truck needs access to the liquid nitrogen reservoir for filling the system.

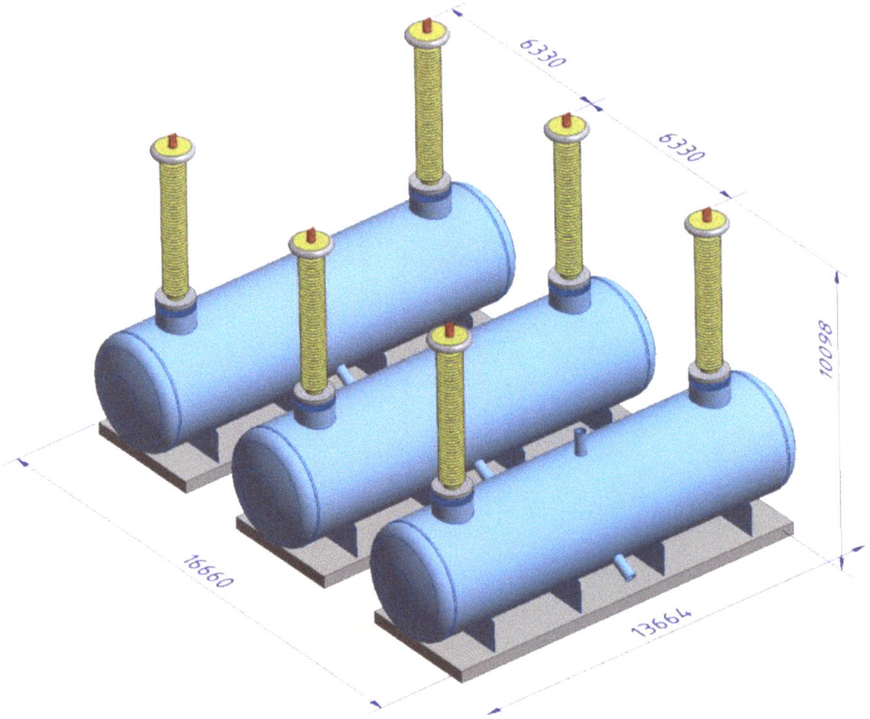

Figure 4.8.3: Sketch of the arrangement of three phase 380 kV SFCL

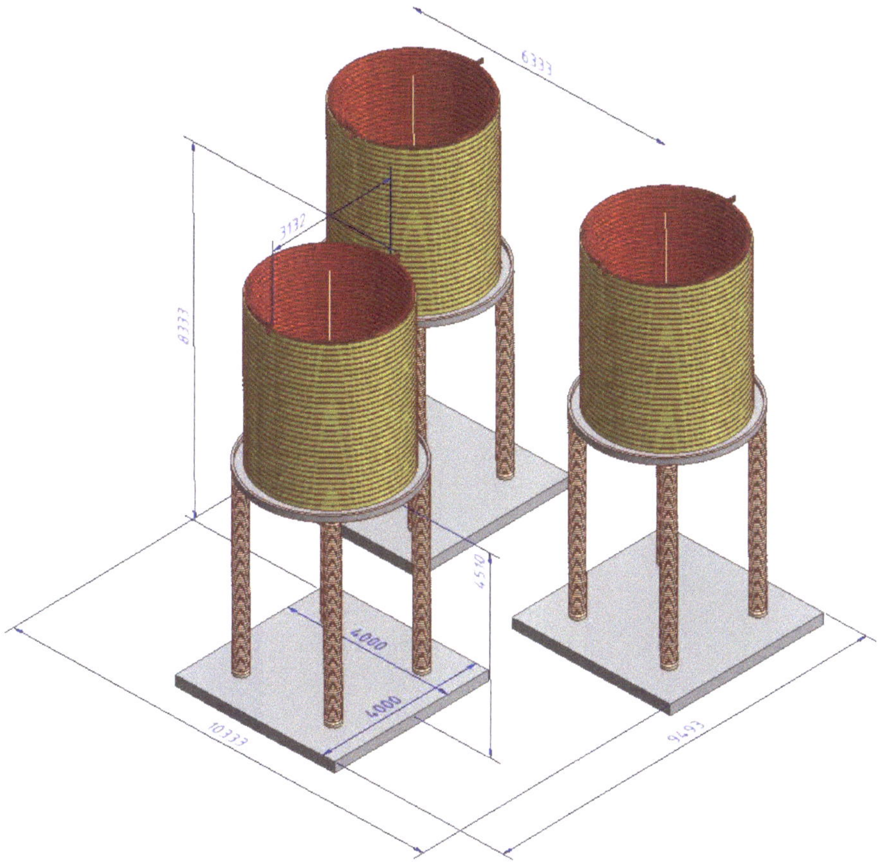

Figure 4.8.4: Sketch of the arrangement of three current limiting reactors connected in series to the 380 kV SFCL

5 Safety and Failures

5.1 Safety Considerations for handling Liquid Nitrogen

Liquid nitrogen is used as a coolant in superconducting cable systems. Liquid nitrogen is easily accessible, inexpensive and environmentally friendly. Some guidelines must be followed when handling this coolant. This chapter summarizes the properties of liquid nitrogen and how to handle it safely.

5.1.1 Properties

Nitrogen (element symbol N) is a chemical element with atomic number 7. It occurs as an element only in the form of diatomic molecules (N_2) and is the main constituent of air (78 %). Molecular nitrogen is a colorless, odorless and tasteless gas, which condenses to a colorless liquid at low temperatures. Nitrogen is sparingly soluble in water and is not flammable. The main physical properties are:

- Density 1.250 kg/m^3 at 273.15 K
- Density 0.809 kg/l (liquid) at 77.15 K (-196 °C), 1 bar
- Melting point 63.05 K (-210.1 °C) at 1 bar
- Boiling point 77.15 K (-196 °C) at 1 bar
- Molar volume (solid) $13.54 \cdot 10^{-6}$ m^3/mol
- Heat of vaporization 5.58 kJ/mol
- Heat of fusion 0.36 kJ/mol
- Speed of sound 333.6 m/s at 298.15 K
- Specific heat capacity 1040 J/(kg·K) at 298 K
- Thermal conductivity 0.02583 W/(m·K)

5.1.2 Hazards

When handling liquid nitrogen, be aware of the following hazards:

Lack of oxygen

Escaping liquid nitrogen evaporates rapidly and cold mist is formed. Evaporation of 1 liter of liquid nitrogen produces approximately 700 liters of gaseous nitrogen. Accumulation of nitrogen in the air reduces the oxygen concentration. At an oxygen content of less than 17 % in the air, people can quickly and without warning become unconscious and suffocate. Enclosed rooms must therefore be equipped with a warning device for oxygen deficiency.

Risk of bursting

The expansion rate of nitrogen from the liquid to the gaseous state is about 1:700 at 1 bar. When liquid nitrogen boils in closed containers, large forces occur which can cause the container to burst.

Oxygen condensation

Vessels containing liquid nitrogen can condense oxygen from the air. Liquid oxygen is a strong oxidizing agent. Organic materials such as oils and greases burn rapidly and strongly in liquid oxygen and can detonate unexpectedly on contact with liquid oxygen.

Cold burns

If liquid nitrogen contacts the skin, frostbite may result. Serious eye damage may occur if liquid nitrogen contacts with eyes.

The following first aid measures should be taken after an accident involving liquid nitrogen:

Wetted clothing, shoes and stockings must be removed immediately. Rinse the affected parts of the body with warm water. Do not rub!

Inhalation

- Bring the injured persons into fresh air using self-contained breathing apparatus.
- Keep the injured persons warm and calm.
- In case of respiratory arrest of the injured persons, ventilation is to be performed.
- A physician shall be consulted.

Ingestion

- Ingestion is not considered a possible route of exposure.

5.1.3 Handling and Protective Measures

There is no hazard from liquid nitrogen during normal operation. For all work where there is a risk of liquid nitrogen being splashed, suitable work clothing or appropriate protective equipment already included in the personal protective equipment of TSO employees must be worn:

- Work clothing
- Safety shoes
- Hand protection
- Eye and face protection
- Possibly respiratory protection

5.2 Failures

Faults and failures can occur in all operating equipment and components and can never be completely ruled out, even with the greatest care. Basically, the significance of the various failures can be assessed with the combination of the consequence, their probability of occurrence and the early and safe detection. However, a detailed "Failure Mode and Effects Analysis" is not the subject of this investigation.

In this study, the faults are distinguished between internal and external faults. Internal faults occur within the current limiter system and external faults affect the current limiter system through external influences. For both types of faults, the consequences and possible remedies are discussed. As far as possible, statements are made on the probability and frequency of occurrence.

Table 5.2.1 summarises the internal faults of the superconducting current limiter, the resulting consequences and remedial measures.

Table 5.2.1: Internal errors of the 380 kV SFCL system with 5 bar cryostat

Description of the error	Consequences	Remedial action
Cryocooler failure	In the open system none In a closed 5 bar system temperature rises » Critical current drops, switch off at max. temperature	Design the cooler redundantly and install it in such a way that it can be changed without shutting down the system.
Failure of the LN_2 pumps	In the 5 bar system Temperature rise in the LN_2 » Critical current drops, switch off at max. temp.	Design pumps redundantly and install them in such a way that they can be changed without shutting down the system.
Vacuum pump failure	Temperature rise in the subcooler causes temperature rise in the LN_2 » Critical current drops, switch off at max. temp.	Design pumps redundantly and install them in such a way that they can be changed without shutting down the system.
Sensor failure	Loss of the respective information of the sensor	Run critical sensors redundantly
Failure of the control and regulation system	No information about the operating status. Can only be tolerated for a short time.	Set up control and regulation twice.
Defect of the cryostat and LN_2 leading lines	Liquid nitrogen discharge » Switch off immediately if necessary	Regular audit.
Degradation of the superconductor	The critical current drops and/or an additional resistance occurs » Switch off and replace affected module	Avoid mechanical or thermal overstressing. Extensive module tests. Regular checking of the critical current density.
Internal short circuit	» Immediate high pressure rise due to arcing in liquid nitrogen. Can lead to destruction of the current limiter module.	Monitoring the quality of the liquid nitrogen. Regular PD measurement.

A <u>cryocooler failure</u> has different effects depending on the cooling system. Since no cryocoolers are required in the open cooling system, there is also no impact in this case. In the closed system at a pressure of 5 bar, a pressure relief valve is also installed that opens slightly above 5 bar. In this case, if the cooler fails, the liquid nitrogen first takes on the temperature of 94 K corresponding to the pressure. This heats up the liquid nitrogen and thus reduces the critical current density. In this case, the current limiting system must be disconnected from the mains at a maximum temperature for safety reasons. Since the cooling system is designed redundantly and a cooler can also be replaced during operation, no influence on the operation of the current limiter is to be expected in the event of a cooler failure.

In the 5 bar subcooled liquid nitrogen system, the failure of the liquid nitrogen pumps initially causes the temperature of the liquid nitrogen in the cryostat to rise, as the further supply of supercooled nitrogen stops. As a result, the critical current in the superconductor decreases and the current limiter must be switched off when the temperature reaches a threshold value. The pumps can be designed redundantly and installed in such a way that, in the event of a pump failure, continued operation is possible and the affected pump can be replaced or serviced separately. Thus, in case of failure of a pump, no influence on the operation of the current limiter is to be expected.

The failure of the vacuum pumps would lead to a rise in temperature in the subcooler and thus the temperature in the cryostat would rise. Similar to the failure of a liquid nitrogen pump, a redundant design means that the failure of a pump has no effect on the operation of the current limiter.

A number of different sensors (e.g. temperature, pressure) are built into the current limiter system to monitor the operating status. A failure of sensors inevitably leads to the loss of the information associated with the sensor and, depending on its importance, to a risk in further operation. Whether this risk can be borne varies depending on the sensor and information. In any case, critical sensors should be redundant. This way, a failure of a sensor would not affect the operation of the current limiter.

A failure of the control and monitoring of the current limiter leads to an immediate loss of information about the operating status. This can usually only be tolerated for a very short time. Without a double design of the control and regulation, the current limiter would have to be switched off instantaneously. As a remedy, a double control and regulation system must be set up. In the event of a failure of one control and regulation system, the other system is then promptly switched over. In this way, this failure should have no effect on the further operation of the current limiter.

A defect of the cryostat or of liquid nitrogen lines usually leads to a leakage of liquid nitrogen and/or to an increased consumption of liquid nitrogen. Depending on the severity of the defect, the current limiter may have to be switched off immediately. A regular visual inspection or thermography of all components carrying liquid nitrogen is recommended as an effective remedy.

Degradation of the superconductor is usually understood to mean a reduction in the critical current density of the superconductor. In this application, however, it can also mean an increase in the contact resistance from the superconductor to the copper. As a result, an additional resistance occurs in the resistive current limiter, which, with good cooling and depending on the length of the affected section, need not yet have any effect on the operation of the limiter. In the worst case, the current limiter must be switched off and the affected current limiter module replaced. A remedy is possible by avoiding mechanical or thermal overstressing of the superconductor during operation. In this case, no degradation of high-temperature superconductors is known so far. Another precaution is that extensive current-limiting tests must be carried out for each module before it is installed in the current limiter. Furthermore, a regular check of the critical current of the limiter is recommended.

The fault with the most severe effects would certainly be an internal short circuit in the cryostat. The associated heat of the arc would briefly generate a large quantity of gas, which could only be discharged by a sufficiently dimensioned rupture disc. The cryostat must be dimensioned in such a way that the associated pressure increase can be handled safely. This fault can lead to a total failure of one phase of a limiter. As a remedial measure, a conservative design for operation with gaseous nitrogen is initially provided and furthermore the voltage tests according to [IEC2019] must be carried out in full. The overpressure in the cryostat prevents air in the cryostat from freezing out and leading to contamination of the liquid nitrogen. This should ensure a high purity of the liquid nitrogen even over a long period of use. Contamination by refilling of liquid nitrogen is prevented by filters.

An overall assessment of the probability of the occurrence of failures and the indication of times such as the mean time between failures is not available due to the small number of current limiters installed so far and the very special design here. Only simple statements about the probability such as high, medium or low would be possible.

Table 5.2.2 summarises the external faults of the superconducting current limiter, the resulting consequences and remedial measures.

Table 5.2.2: External errors

Description of the error	Consequences	Remedial action
Automatic reclosing	None	None
External overtemperature	Increase the heat input into the cryostat. Shorter refill times Lowering the LN_2 level	If critical, then increase cooling capacity
Temporary overvoltages and overcurrents	None, if considered in the design	Surge arrester at both ends
Failure of the external circuit breaker	None, as considered in the fault durations	None
Lightning strike	None, as considered in the design	Sufficiently dimension the earthing system and provide suitable lightning protection measures.
Power supply failure	Without UPS, causes the current limiter to switch off	Provide emergency power supply if necessary
Human error	Very diverse. E.g. Minimum LN_2 level is not reached.	Automatic emergency shutdowns in case of operating errors
Extreme weather influences	Can lead to total loss of the current limiter	Design for maximum events, e.g. ice load, rain, etc.

Automatic reclosure or short interruption is used to extinguish non-permanent arcs caused, for example, by fallen branches on overhead lines without permanently interrupting the supply. In this case, the responsible circuit breaker trips by energising the line protection and automatically closes again after a short time. Whether a superconducting current limiter must be disconnected from the mains in this case depends on the earthing of the mains and the type of fault (3-phase symmetrical or asymmetrical) and cannot be answered in general terms. In general, a fault-ride-through capability with a pause time can be taken into account in the design and thus the current limiter can remain connected to the mains after the automatic reconnection occurs. This design option is not considered in this study but a positive experience was also had with the 10 kV, 40 MVA limiter in the AmpaCity project in Essen. There the limiter was not disconnected after an automatic reclosure. Since the earthing of the grids and the effect of unbalanced faults are not known and are not part of these investigations, no detailed statements can be made on this.

An external over-temperature outside the maximum permissible external temperature leads to an increased heat input into the cryostat of the current limiter. This increase in heat input can be easily compensated for open cooling by shorter refilling times. With closed cooling, in the worst case, the cooling capacity would no longer be sufficient and the limiter would have to be switched off. In any case, this can be easily detected by the existing temperature measurement in the liquid nitrogen and a reaction can be made in time. An additional effort for handling this error is not to be expected.

Temporary overvoltages and overcurrents are to be specified in a detailed specification according to [CIGRE2012] and taken into account accordingly when designing the current limiter system. In this study, the design is based on the maximum voltage and the rated current. Should a limitation occur in one phase of a temporary overvoltage, this would possibly lead to a higher load on the superconductor. This can be taken into account accordingly in the design. A temporary overcurrent leads to higher AC losses and, depending on the level, can also cause the current limiter to trip. Therefore, knowledge of the temporary overcurrents and the minimum short-circuit currents is required for precise coordination of the limiting behaviour. A permanent, safe operation of the limiter is safely guaranteed up to the critical current and thus the overcurrents need to be known and considered in the design.

A failure of the external circuit breaker has no effect on the safety of the SFCL, as this case has already been taken into account in the maximum fault durations. It is assumed that the tripping of a back-up protection is included in the mentioned fault duration.

In the event of a lightning strike, the main tasks of the earthing system are to discharge the lightning current and to avoid dangerous overvoltages. To achieve these goals, the earthing system must be sufficiently dimensioned. The lightning protection measures are contained in DIN VDE 0101-1 [DIN2014] and the procedure to be used must be agreed between the manufacturer of the current limiting system and the operator. To limit overvoltages in the current limiter, surge arresters are provided on both terminals. For liquid nitrogen tanks, the technical guideline [IGV2018] of the Industrial Gases Association for earthing, equipotential bonding and lightning protection on tank installations must also be observed.

In the case of a lightning strike, a distinction can be made between an impact on an overhead line near the limiter and an impact directly into the cryostat. A lightning strike on an overhead line results in an incoming travelling wave, which should be safely handled by the surge arresters and the design for the lightning impulse voltage of 1425 kV (LI) and 1570 kV (LIC).

What happens in the event of a lightning strike to the cryostat if the lightning strikes the outer vessel directly? This is not a pressure-bearing part, but can be understood to be a protective shell of the inner, actual pressure vessel. The current is discharged into the ground via the container and its earthing or - if present - via the equipotential bonding. The latter is not even a fixed requirement for liquid nitrogen containers. As the walls of the containers are relatively thick, damage to the cryostat is not to be expected and an inspection is not necessary. The cryostat would only have to be inspected for damage if burn marks on the paint or deformations on the walls were visible. In fact, however, a lightning strike cannot be visually detected, even if it has occurred. Practical experience shows that lightning strikes occasionally occur on liquid nitrogen tanks. Since telemetry systems are installed on the tanks and can be damaged by lightning (overvoltage), it is known that some tanks are affected by lightning each year.

A <u>failure of the power supply for the auxiliary energy</u> to operate the limiter leads to different effects depending on the cooling variant. At 5 bar in the open system, an uninterruptible power supply has to operate the liquid nitrogen pumps and the vacuum pumps over the bridging period. This power of the uninterruptible power supply would increase even more in the closed system, as the cryocoolers and their compressors would now have to be supplied. Overall, a failure of the power supply does not have to lead to a shutdown of the current limiter system, but depending on the cooling variant, uninterruptible power supplies of different dimensions must be provided.

<u>Human error</u> cannot be completely ruled out even if the operating personnel are well qualified and can essentially occur during operation and maintenance of the current limiter. In principle, faulty operation must be prevented by protective circuits and interlocks. Switching on the current limiter when the liquid nitrogen level is not yet sufficient must be prevented, e.g. by not enabling the switch-on. Whether this is to be cancelled by the operator in an emergency must be clarified in the integration of the current limiter into the operator's monitoring and protection system. If the liquid nitrogen level drops below the minimum value during operation, this must be indicated in good time by warnings and finally switched off automatically, as continued operation would lead to permanent damage to the current limiter. Errors in maintenance measures can be detected by a current and voltage test before switching on again. Whether this is desired must be agreed individually with the respective operator.

To minimise the probability of human error, a clear traffic light system with green for OK, yellow as a warning level and red as an alarm level for the components of the current limiter system is recommended for communication with the operator's operating system. However, since the integration must be carried out in detail with the respective operator, this is not discussed further in this study.

A <u>failure of the circuit breaker</u> to clear the fault in the network should not have any influence, as this case should already be considered in the fault durations.

Extreme weather influences such as those caused by storms, hail, floods, snow, freezing rain or similar can be considered through a detailed risk assessment and partly taken into account in the design. For example, the maximum amount of rain, the maximum wind speed, the maximum snow and ice load and the corresponding earthquake zone would have to be taken into account in the design at the respective installation site. A current limiting system could be designed for this and could also continue to operate when it occurs.

6 Test Protocols and Laboratories

6.1 Applicable Standards

Since 2015, the general IEEE test recommendation "IEEE Guide for Fault Current Limiter (FCL) Testing of FCLs Rated above 1000 V AC" [IEEE2015] is available for most types of fault current limitation systems. These include reactors, circuit breakers, I_S-limiters as well as resistive and inductive superconductive fault current limiters. Within this test guide, a specific test sequence is defined, but essentially reference is made to known test standards for conventional electrical equipment, such as transformers, circuit breakers, switchgear, etc. The standards that are relevant for fault current limiter are listed in Table 6.1.1 according to [IEEE2015]. In addition to the electrical test specifications, the standard [IEEE2015] includes further specifications for physical and operational parameters, environmental aspects, safety and lifetime. In the following, only the direct and indirect electrical test specifications are dealt with. These are also applied here to the SFCL system. For further and more detailed information regarding the performance, evaluation, etc. of the individual tests, please refer to [IEEE2015] as well as the respective relevant standard (Table 6.1.1).

Table 6.1.1: Extraction of specified test standards according to [IEEE2015]

Number of Standard	Title of Standard
IEC 60060-1	High-voltage test techniques – Part 1: General definitions and test requirements
IEC 60071-1	Insulation co-ordination – Part 1: Definitions, principles and rules
IEC 60076-3	Power transformers – Part 3: Insulation levels, dielectric tests and external clearances in air
IEC 60076-4	Power transformers – Part 4: Guide to the lightning impulse and switching impulse testing Power transformers and reactors
IEC 60076-10	Power transformers – Part 10: Determination of sound levels
IEC 60137	Insulated bushings for alternating voltages above 1.000 V
IEC 60255-26	Measuring relays and protection equipment – Part 26: Electromagnetic compatibility requirements
IEC 60270	High-voltage test techniques – Partial discharge measurements
IEC 62271-1	High-voltage switchgear and controlgear – Part 1: Common specifications for alternating current switchgear and controlgear
IEC 62271-100	High-voltage switchgear and controlgear – Part 100: Alternating-current circuit-breakers

To check the electrical performance in general, the standard [IEEE2015] recommends (if possible and reliable) performing the following tests as part of a design test:

- **High voltage tests**
  - Power frequency withstand voltage test
  - Impulse voltage test
  - Partial discharge level test
- **High current tests**
  - Rated continuous current test

    - Short-time withstand and peak withstand current test
    - Short-circuit current limitation test
    - Recovery test (Transient recovery voltage test)
- **Combined power test**
- **Electromagnetic compatibility test**
- **Seismic test**
- **Audible sound test**

The examinations listed do not represent an order or repetition. Which test should be carried out or repeated and when is still to be developed, especially for a resistive superconducting fault current limiter. In the following, depending on the standards mentioned in Table 6.1.1 and the high voltage requirements in Table 2.2.2, the test procedures for the tests mentioned in the previous listing are explained.

6.1.1 Power Frequency Withstand Voltage Test

For the power frequency withstand voltage test of a 420 kV (U_{m}) device, an effective test voltage of 630 kV with a frequency of 50 Hz must be applied for a duration of 60 s in accordance with IEC 60076-3, which refers to the standardised test levels of IEC 60071-1. There must be no breakdown.

6.1.2 Impulse Voltage Test

During the impulse voltage test, different impulse tests must be carried out depending on the highest voltage. The mentioned test standard [IEEE2015] refers to IEC 60076-3, which in turn refers in parts to IEC 60076-4 and IEC 60060-1. Up to a maximum voltage of 245 kV, only a lightning impulse voltage test must be carried out. Above this value, a switching impulse voltage test and, especially from the area of transformer testing, a chopped lightning impulse voltage test are added. According to [IEEE2015], it is therefore proposed to carry out all three tests for the SFCL here. However, it should be mentioned that the referenced standards IEC 60076-3, IEC 60076-4 and IEC 60060-1 are quite different with regard to the execution of the impulse voltage tests. In the more general IEC 60060-1, the individual impulse tests are defined independently of each other. While IEC 60076-4, which applies specifically to impulse tests on transformers, defines combined impulse test procedures. It should also be noted that IEC 60060-1 is superior to IEC 60076-3 and IEC 60076-4. Both also refer in essential parts to the specifications of IEC 60060-1. Whether the procedures of IEC 60076-3 and IEC 60076-4 are the most suitable procedures for impulse testing cannot be clarified at this point. However, they will be explained in the following.

6.1.2.1 Combined Full and Chopped Lightning Impulse Voltage Test

The full lightning impulse is defined with a front time of 1.2 µs (± 30 %), a time to half value of 50 µs (± 50 %) and an amplitude of 1425 kV (± 3 %). The chopped lightning impulse is also defined with a front time of 1.2 µs (± 30 %) but is chopped in the back within a cut-off time of 2 to 5 µs. The amplitude is 1570 kV (± 3 %) with a factor of 1.1 of the amplitude of the full lightning impulse. In general, both impulses stresses can be generated or demanded with positive as well as negative polarity. The IEC 60076-4 standard only recommends carrying out with a negative polarity, as this considerably reduces the probability of an external flashover.

The combined full and chopped lightning impulse voltage test according to IEC 60076-4 requires the following test sequence:

1. 1x lightning impulse with 50 – 75 % of max. amplitude
2. 1x lightning impulse with 100 % of max. amplitude
3. Min. 1x chopped lightning impulse with less than 100 % of max. amplitude
4. 2x chopped lightning impulse with 100 % of max. amplitude
5. 2x lightning impulse with 100 % of max. amplitude

The test is considered passed if there is no breakdown of the tested insulation system.

6.1.2.2 Switching Impulse Voltage Test

In general, a switching impulse voltage is defined with a front time of 250 µs (± 20 %) and a tail time of 2500 µs (± 60 %). The amplitude is 1175 kV (± 3 %). The polarity is as described in section 6.1.2.1. Although IEC 60076-3 refers to IEC 60076-4 for the impulse voltage tests, the latter in turn refers back to IEC 60076-3 for the execution of switching impulse voltages.

The switching impulse voltage test according to IEC 60076-3 requires the following test sequence:

1. 1x switching impulse with 50 – 70 % of max. amplitude
2. 3x switching impulse with 100 % of max. amplitude

The test is considered passed if there is no breakdown of the tested insulation system.

6.1.3 Partial Discharge Level Test

When performing the partial discharge test, the standard [IEEE2015] opens up two different procedures with different test times, test voltages and partial discharge levels as test criteria by naming the standards for electrical bushings IEC 60137 and transformers IEC 60076-3. Both refer to the general standard for partial discharge measurements IEC 60270. Since IEC 60076-3 is an induced AC test with partial discharge measurement, a partial discharge measurement according to IEC 60137 is considered more suitable as a test for a resistive superconducting current limiter. Since the maximum test voltage for the partial discharge test is based on the test voltage of the AC power frequency withstand voltage test according to IEC 60071-1 (see section 6.1.1), the two tests can be combined. Otherwise, the test voltage can also be applied again.

The partial discharge level test according to IEC 60137 requires the following test sequence:

1. Start up the AC power frequency withstand voltage and hold for 5 minutes
2. Continuous lowering of the test voltage
3. From U_{m} (as phase-earth voltage) at the latest, the apparent charge must be less than 10 pC

Whether the test voltages and partial discharge limits provide a valid statement about the insulation system of superconducting equipment is not proven according to the current state of the art. There is currently no research knowledge on how partial discharges under cryogenic conditions contribute to the electrical ageing of insulation systems. Conversely, it is also not currently known whether cryogenic insulation systems can withstand higher partial discharge stresses.

6.1.4 Rated Continuous Current Test

In the rated continuous current test in accordance with IEC 62271-1, the functionality of the superconducting current limiter is tested under rated current conditions. The test current is successively increased and key parameters defined by the manufacturer are recorded and evaluated for their steady-state conditions. The test is successfully completed when the rated current, in the cases A and B considered here, is 5 kA at a frequency of 50 Hz and the key parameters have reached their expected steady state.

6.1.5 Short-time Withstand and Peak Withstand Current Test

The short-time and peak withstand current test according to IEC 62271-1 serves to verify that the superconducting current limiter is capable of carrying its rated short-time and peak withstand current during the rated short-circuit duration. For case A and case B, the rated short-time short-circuit currents are 19 kA and 30 kA. The rated peak withstand current for both cases is 63 kA and the rated short-circuit duration is 0.25 s. The test only needs to be performed once and is essentially considered a pass if there are no unexpected deviations. If eligible, this examination may be combined with the examination in section 6.1.4.

6.1.6 Short-Circuit Current Limitation Test

In the short-circuit limitation test in accordance with IEC 62271-100, the electrical, thermal and mechanical behaviour of the current limiter caused by a limiting event is tested. The test should (as far as possible) be carried out under rated conditions. This means with a rated current of 5 kA, a rated voltage of 420 kV and a rated frequency of 50 Hz. In addition, the prospective minimum and maximum peak withstand short-circuit currents with the fault duration of 0.25 s should be carried out. At least one test shall be carried out for each of the minimum and maximum peak withstand short-circuit currents and for different polarities (if different effects are expected). If the test laboratory is not able to implement these due to the high rating requirements, they may be adjusted appropriately.

The short-circuit current limitation test according to [IEEE2015] requires the following test sequence:

1. Rated current for min. three seconds
2. 10 repetitions of the prospective peak withstand short-circuit current
3. Rated current for twice the expected recovery time

If there is no deviation in nominal operation after the final nominal stressing, the test is considered to have been passed. If deemed appropriate, the examination here may also be combined with those from section 6.1.4 and section 6.1.5.

6.1.7 Electromagnetic Compatibility Test

For electromagnetic compatibility testing, [IEEE2015] refers to IEC 62271-1. According to this, a distinction must be made between two tests. On the one hand, the emission tests on main circuits (radio interference voltage tests) and, on the other hand, the test of interference emissions from auxiliary and control circuits. In the following, only the former will be discussed. The radio interference voltage test is carried out on the complete equipment including all standard attachments. According to IEC 62271-1, the test setup and the test procedure must comply with CISPR TR 18-2 [CISPR2017]. Then, a voltage of $1.1\ U_{\mathrm{m}}\ /\sqrt{3}$ is applied for five minutes.

The test is passed if the radio interference voltage measured value does not exceed the radio interference level of 2500 µV.

6.1.8 Seismic Test

For seismic testing, [IEEE2015] refers to IEEE Standard 693. This standard, which is also referenced by IEC 62271-1, provides tests and evaluation procedures for seismic testing. In order to define the type of stress that may occur, a complex risk analysis must be carried out, which is essentially dependent on the installation site. This then results in whether a low, moderate or high seismic qualification must be carried out. The equipment-specific characteristics are also to be taken from IEEE Standard 693 in the respective equipment category. The qualification can be carried out by three methods, depending on the previous results. If the equipment is, so called in the standard, “marginally inherently acceptable”, only the anchorage calculation has to be carried out. Otherwise, either a practical equipment qualification in a suitable testing laboratory or, if permissible according to IEEE Standard 693, an analytical equipment qualification must be carried out. As a result of the two possible test qualifications, the necessary site and installation conditions, like anchorage specifications, site influences, necessary seismic protection devices, etc. for safe operation of the equipment follow from the seismic point of view.

6.1.9 Audible Sound Test

The standard [IEEE2015] refers to the NEMA standard 20 "General applications for dry-type transformers" for the audible sound test. This standard specifies a maximum permissible average noise level of 67 decibels (dB(A)), depending on the power category up to 1 MVA. During the measurement, the equipment must be arranged in such a way that as few acoustic reflections as possible distort the measurement. The average noise level is determined by taking several measurement points around the equipment in defined distance depending on the equipment height.

6.1.10 Conclusion

The test guide [IEEE2015] provides a good basis for a superconducting fault current limiter to get an overview of necessary and relevant tests and their standards. However, this guide [IEEE2015] applies to a wide range of current limiting technologies of various types. This can lead to a situation where various tests provide a good approach but do not yet represent the optimum for testing a superconducting resistive current limiter. An appropriate solution for this circumstance cannot be worked out within the scope of this study. Therefore, it is necessary for the future that a specific test standard is developed especially for superconducting electrical high-voltage equipment, in this case resistive current limiters.

Furthermore, the tests and their test criteria based on the test guide [IEEE2015] and the standards in Table 6.1.1 are understandable as minimum requirements and may vary in consultation with a customer.

6.2 Test Laboratories

There are two categories of test laboratories. Firstly, there are laboratories in which individual electrical components, such as tape conductors, insulators, modules, bushings, etc., are examined experimentally in liquid nitrogen with regard to their basic electro-technical function. Secondly, there are accredited test laboratories where the entire electrical system of the SFCL can be tested according to normative type test specifications. Both types of laboratories will be necessary, as it is currently not possible to test the electrical insulation of the bifilar-wound modules by means of a partial discharge measurement, for example, when testing the assembled overall system. This must take place in a preliminary test on the manufactured subcomponents.

6.2.1 Experimental Test Laboratories

In addition to high-voltage and high-current equipment, the experimental test laboratories (Table 6.2.1) also have specific cryogenic test equipment for the experimental investigation of superconducting equipment and subcomponents. This essentially includes the liquid nitrogen bath cryostat as well as the associated equipment and the entire infrastructure for carrying out tests under high-voltage stress inside the cryostat.

The experimental facilities listed in Table 6.2.1 are not suitable for the rated voltage of 420 kV in order to carry out investigations on fully scaled components. This would require larger cryostats with higher test voltages, which do not yet exist. However, investigations on downscaled experimental components are possible. Table 6.2.1 includes all test laboratories located in Europe and their cryogenic experimental facilities.

Table 6.2.1: Overview of the cryogenic experimental facilities located in Europe [CIGRE2021]

Place	Country	$V_{AC,RMS}$	$V_{LI,PEAK}$	Maximum pressure p_{ABS}	Cryostat dimension
Calais	France	232 kV	±750 kV	0.1 MPa	h: 2200 mm, d: 550 mm
Grenoble	France	100 kV	200 kV	0.5 MPa	h: 300 mm. d: 400 mm
Lyon	France	100 kV	-	0.1 MPa	1.5 × 1.5 × 0.6 m^3
Karlsruhe	Germany	230 kV	±365 kV	0.3 MPa	h: 1016 mm. d: 642 mm
Cologne	Germany	230 kV	±605 kV	0.5 MPa	h: 1200 mm. d: 842 mm
Braunschweig	Germany	120 kV	±80 kV	0.3 MPa	h: 200 mm, d: 394 mm
Hannover	Germany	350 kV	1000 kV	-	h: 1050 mm, d: 590 mm
Milan	Italy	50 kV	125 kV	0.3 MPa	h: 2000 mm, d: 900 mm

6.2.2 Type Test Laboratories

In order to type-test high-voltage equipment systematically according to the standardized requirements mentioned in section 6.1, there are only a few laboratories in Europe that fulfil the high-voltage and high-current requirements. The suitable laboratories with the corresponding testing facilities are the IPH in Berlin and the FGH in Mannheim (Germany), CESI in Milan (Italy), KEMA in Arnhem (Netherlands) and ETC in Prague (Czech Republic). All these laboratories belong to the CESI group, which has its headquarters in Milan.

In addition to conducting standardised type tests, the CESI group would hypothetically also be the only laboratory suitable for conducting a combined performance test as proposed in section 6.1.6. However, the technical requirements for the test systems must first be evaluated and test procedures and test scenarios must be developed in order to be able to carry out this specific performance test. Therefore, no further statements on the feasibility of a performance test according to section 6.1.6 can be made within the scope of this study. However, it is recommended that such a test procedure be developed.

7 Economic Feasibility

7.1 Estimation of CAPEX

The cost estimation is based on industry feedback for the major cost components and an estimate of resources to manufacture and test the fault current limiter. The estimation of the resources was checked with two independent interviews with experts in manufacturing superconducting devices. The following items are excluded from this cost estimation:

- Foundations and earthworks
- Barriers and fencing
- Access controls
- Site safety
- Connections to 380 kV
- Other 380 kV components than SFCL cryostat and shunt reactor
- Permits of any kind
- Commercial conditions
- Risks, warranty
- R&D costs refunds

This cost estimate includes all materials and components of a three phase SFCL system, the main work for construction, manufacturing, assembly and testing.

7.1.1 Assumptions

The main parameters used for the investigation of the economic feasibility are summarized in Table 7.1.1. A detailed sensitivity analysis was not within the scope of this study. Nevertheless, the influence of the load factor on the operation cost was shown later in this chapter. This is to show the influence of the current on the operation cost. For the manufacturing of the current limiting modules and the complete cryostat assembly, the working hours were estimated and the cost was calculated using the manpower costs given in Table 7.1.2. The manufacturing effort and the respective cost was determined after two expert interviews made with senior experts from HTS and manufacturing industry.

Table 7.1.1: Main data used for investigation of economic feasibility

Description	**Data**
Electricity cost	0.293 €/kWh
Rated current	5000 A
LN_2 evaporation rate at 77 K, 1 bar	20 kg/kWh
Liquid nitrogen cost	0.3 €/liter
Hours per year	8760 h
Load factor	0.70
Work loss factor	0.62
Work loss factor HTS	0.38
Electric power for pumps total	2.00 kW
Cryocooler efficiency	0.066
WACC	4.5 %
Use Time	40 years

Table 7.1.2: Data used for estimating labour cost for manufacturing of SFCL

Qualification	Cost
„Facharbeiter" Craftsman	75 €/h
Technician	92 €/h
Engineer	105 €/h

7.1.2 HTS Material Cost

A minimum set of parameters was developed to specify the HTS material for the superconducting fault current limiter. These parameters include:

- Average critical current per tape at 77 K, self field: 700 A
- Tape width: 10-12 mm
- Single piece length: 39 m Case A, 32 m Case B
- Single piece length must be this length or two or three times this value
- Cap layer: Ag 2 µm each side, no copper or other layers
- Total tape amount: 210 km Case A, 240 km Case B

This information was sent to the companies listed in Table 7.1.3 with a set of the following questions:

- Can you deliver this specification: yes or no?
- If yes, what is the cost per km tape?
- If yes, what would be the delivery time?

Table 7.1.3 summarizes the feedback of the companies. All companies claimed that they can fulfil the specification. The cost of the HTS tape covers a wide range from 38 €/m up to 94 €/m and is the most dominant cost share of the fault current limiter. For the medium cost a value of 50 €/m was assumed in this study.

Table 7.1.3: Summary of HTS material information based on feedback from manufacturers

Company	Fulfil specification	Cost of HTS tape in k€/km	Delivery time /months
A	Yes	42.0	12
B	Yes	54.1	12
C	Yes	46.0-94.0	6
D	Yes	38.1-57.1	12
E	Yes	-	-

7.1.3 HTS Current Limiting Element

Table 7.1.4 summarizes the effort to manufacture the current limiting element specified in chapter 4.3 for case A and Table 7.1.5 for case B. Three major cost items were identified. These are the HTS material, the cost for the different components of the HTS modules and the work hours for manufacturing. Due to the preliminary nature of the study, a fixed cost based on offers from industry was not available. Therefore, a range with minimum and maximum cost for the specific items was given.

Table 7.1.4: Cost for manufacturing the current limiting element for case A (annotations see Table 7.1.5)

	Min. spec. cost	Med. spec. cost	Max. spec. cost	Amount	Min. cost / k€	Med. cost / k€	Max. cost / k€
HTS element					**9817.204**	**14885.174**	**24892.546**
HTS material	30 €/m	50 €/m	90 €/m	246960 m	**7408.800**	**12348.000**	**22226.400**
FCL module					**1637.332**	**1701.817**	**1766.303**
FCL module invest					**770.872**	**835.357**	**899.843**
Shielding rings Al [1)]	30 €	35 €	40 €	285	130.412	152.147	173.883
Kapton	0.50 €/m	0.50 €/m	0.50 €/m	246960 m	123.480	123.480	123.480
Corrugated insulation	0.50 €/m	0.50 €/m	0.50 €/m	246960 m	123.480	123.480	123.480
Copper [2)]	300 €	350 €	400 €	285	171.000	199.500	228.000
GRP plates [3)]	250 €	275 €	300 €	570	142.500	156.750	171.000
Manufacturing place FCL module	30 k€	30 k€	30 k€	1	30.000	30.000	30.000
Manufacturing place isolated twins	50 k€	50 k€	50 k€	1	50.000	50.000	50.000
FCL module work					**866.460**	**866.460**	**866.460**
Manufacturing of isolated twins	8 h	8 h	8 h	285	190.380	190.380	190.380
Manufacturing of modules	8 h	8 h	8 h	285	190.380	190.380	190.380
Module current test	8 h	8 h	8 h	285	224.580	224.580	224.580
Total mod. assembly	960 h	960 h	960 h	3	261.120	261.120	261.120

Table 7.1.5: Cost for manufacturing the current limiting element for case B

	Min. spec. cost	Med. spec. cost	Max. spec. cost	Amount	Min. cost / k€	Med. cost / k€	Max. cost / k€
HTS element					**9065.784**	**13596.715**	**22518.747**
HTS material	30 €/m	50 €/m	90 €/m	219555 m	**6586.650**	**10977.750**	**19759.950**
FCL module					**1698.285**	**1768.201**	**1838.116**
FCL module invest					**780.849**	**850.765**	**920.680**
Shielding rings Al [1)]	30 €	35 €	40 €	309	141.394	164.960	188.525
Kapton	0.50 €/m	0.50 €/m	0.50 €/m	219555 m	109.778	109.778	109.778
Corrugated insulation	0.50 €/m	0.50 €/m	0.50 €/m	219555 m	109.778	109.778	109.778
Copper [2)]	300 €	350 €	400 €	309	185.400	218.300	247.200
GRP plates [3)]	250 €	275 €	300 €	618	154.500	169.950	185.400
Manufacturing place FCL module	30 k€	30 k€	30 k€	1	30.000	30.000	30.000
Manufacturing place isolated twins	50 k€	50 k€	50 k€	1	50.000	50.000	50.000
FCL module work					**917.436**	**917.436**	**917.436**
Manufacturing of isolated twins	8 h	8 h	8 h	309	206.412	206.412	206.412
Manufacturing of modules	8 h	8 h	8 h	309	206.412	206.412	206.412
Module current test	8 h	8 h	8 h	309	243.492	243.492	243.492
Total mod. assembly	960 h	960 h	960 h	3	261.120	261.120	261.120

1) 30 € per kg

2) 300 € per module for 12 contacts plus inner ring for connection

3) 100 € per plate plus 150 € for cutting

It can be seen that the largest cost share is the HTS material cost with also the largest uncertainty. Minimum and maximum cost for the HTS material was taken from Table 7.1.3.

7.1.4 Cryostat and Bushing

Two companies that provide bulk liquid nitrogen containers were asked if they can manufacture a large liquid nitrogen cryostat as specified in chapter 4.8.2. They made a cost estimate range between 1 Mio. € and 1.4 Mio. € for one cryostat. In addition, the cost for the filling of the cryostat with appr. 120.000 liters of liquid nitrogen was assumed.

For the cost of the bushing an estimate from a company was taken. This ranged from appr. 200 k€ up to 280 k€ per 380 kV bushing.

7.1.5 Cooling System

For the open cooling system described in chapter 4.7.2 a cost estimate was given from an industry expert. Depending on the detailed specification the system cost ranges between 2.2 Mio. € and 2.35 Mio. €. This number is without a liquid nitrogen reservoir that is usually provided with the delivery contract for liquid nitrogen. The cryocooler cost is assumed within a range of 75 and 125 € per Watt of cooling power at 70 K. The respective cooling power is given in chapter 4.6.5. Additionally, the cost for the measurement and control system (MSR) was estimated with a best guess approach. The total cost for the cooling system is summarized in Table 7.1.6.

Table 7.1.6: Cost estimate for the SFCL cooling system

	Min. cost	Med. cost	Max. cost
	3385 k€	**3939 k€**	**4325 k€**
Open cooling system	2200 k€	2350 k€	2350 k€
Cryocooler for closed system	1185 k€	1580 k€	1975 k€
MSR	150 k€	175 k€	200 k€

Since the maximum heat input to low temperatures is very similar for case A and B the same cooling cost is assumed in both cases.

7.1.6 Current Limiting Reactor

Two companies were asked for the cost of the conventional current limiting reactor. Their feedback is summarized in Table 7.1.6. It should be noted that the current limiting reactors are not suitable to carry the rated current continuously. During normal operation, the current in these reactors can be neglected due to the parallel connection of the superconductor.

Table 7.1.7: Summary of investment cost of conventional current limiting reactor

	Manufacturer A		Manufacturer B	
Cost	**Case A**	**Case B**	**Case A**	**Case B**
Reactor	150 k€	170 k€	-	-
Substructure @ 0.3g	100 k€		-	-
Substructure @ 0.6g	150 k€		-	-
One phase @ 0.0g	-		400 k€	550 k€
One phase @ 0.3g	250 k€	270 k€	-	-
One phase @ 0.6g	300 k€	320 k€	-	-
Three phase system @ 0.0 g	**-**	**-**	**1200 k€**	**1650 k€**
Three phase system @ 0.3 g	**750 k€**	**810 k€**	**-**	**-**
Three phase system @ 0.6 g	**900 k€**	**960 k€**	**-**	**-**

7.1.7 Others

The cost items in this section (see Table 7.1.8) include the manpower for project management (one engineer for 3 years of development), construction drawings (two engineers, each for 1.5 years), documentation (1 engineer for one year) and transport. Based on specific safety requirements and transport distance, the effort for transport and documentation can deviate significantly.

Table 7.1.8: Cost estimate for other cost parts

	Min. cost	Med. cost	Max. cost
Total	**1175.2 k€**	**1225.2 k€**	**1275.2 k€**
Construction	336.0 k€	336.0 k€	336.0 k€
Documents	184.8 k€	184.8 k€	184.8 k€
Transport	100.0 k€	150.0 k€	200.0 k€
Project management	554.4 k€	554.4 k€	554.4 k€

7.1.8 Summary

Based on the numbers given in the previous sub-chapters, the total CAPEX for a three phase 380 kV SFCL is given in Table 7.1.9 for case A and Table 7.1.10 for case B. It can be seen that the largest part and the largest uncertainty comes from the HTS material. This can be overcome only by asking for detailed offers and delivery times, which was not within the scope of this study. The CAPEX for a three phase 380 kV SFCL system ranges from 19 Mio. € to 37.5 Mio. € for case A and from 18.2 Mio. € to 35.1 Mio. € for case B.

Table 7.1.9: Summary of investment cost of major cost components for fault current limiter case A

	Minimum	Medium	Maximum
Total invest cost	**19011.4 k€**	**25805.6 k€**	**37538.0 k€**
High temperature superconductor	7408.8 k€	12348.0 k€	22226.4 k€
FCL module	1637.4 k€	1701.9 k€	1766.4 k€
- FCL module invest	770.9 k€	835.4 k€	899.9 k€
- FCL module work	866.5 k€	866.5 k€	866.5 k€
Cryostat	3043.0 k€	3647.5 k€	4252.0 k€
Bushing	1213.0 k€	1453.0 k€	1693.0 k€
Cooling & MSR	3385.0 k€	3930.0 k€	4325.0 k€
System assembly and tests	250.0 k€	300.0 k€	350.0 k€
Others, transport	1175.2 k€	1225.2 k€	1275.2 k€
Shunt Coil	900.0 k€	1200.0 k€	1650.0 k€

Table 7.1.10: Summary of investment cost of major cost components for fault current limiter case B

	Minimum	Medium	Maximum
Total invest cost	**18251.4 k€**	**24502.0 k€**	**35143.4 k€**
High temperature superconductor	6586.7 k€	10977.8 k€	19760.0 k€
FCL module	1698.4 k€	1768.3 k€	1838.2 k€
- FCL module invest	780.9 k€	850.8 k€	920.7 k€
- FCL module work	917.5 k€	917.5 k€	917.5 k€
Cryostat	3043.2 k€	3647.6 k€	4251.9 k€
Bushing	1213.1 k€	1453.1 k€	1693.1 k€
Cooling & MSR	3385.0 k€	3930.0 k€	4325.0 k€
System assembly and tests	250.0 k€	300.0 k€	350.0 k€
Others, transport	1175.2 k€	1225.2 k€	1275.2 k€
Shunt Coil	900.0 k€	1200.0 k€	1650.0 k€

7.2 Estimation of OPEX

To calculate the operation cost of the SFCL the loss energy for the specific items was calculated first and is given in Table 7.2.1 for case A and Table 7.2.2 for case B. It assumes that this energy is provided by electricity with the cost given in Table 7.1.1 and that we have a closed cooling system. It can be seen that the load factor has a significant influence on the operation cost. The much higher AC loss with a higher load factor leads to nearly twice the operation cost between a load factor of 0.7 and 1. In summary, the annual operation cost varies between 315 k€ and 695 k€ for case A and 289 k€ and 628 k€ for case B.

Table 7.2.1: Losses and related operation cost per year for case A

	Value	**Value**
Load factor	**0.7**	**1**
AC loss	786.46 MWh	2057.01 MWh
Current lead thermal	103.93 MWh	103.93 MWh
Current lead ohmic	40.07 MWh	65.02 MWh
Cryostat	119.45 MWh	119.45 MWh
Others	7.96 MWh	7.96 MWh
Pumps	17.52 MWh	17.52 MWh
Total energy loss per year	1075.40 MWh	2370.89 MWh
Cost per year for losses	315.450 k€	695.462 k€

Table 7.2.2: Losses and related operation cost per year for case B

	Value	**Value**
Load factor	**0.7**	**1**
AC loss	699.23 MWh	1828.85 MWh
Current lead thermal	103.93 MWh	103.93 MWh
Current lead ohmic	40.07 MWh	65.02 MWh
Cryostat	103.93 MWh	103.93 MWh
Others	7.96 MWh	7.96 MWh
Pumps	17.52 MWh	17.52 MWh
Total energy loss per year	988.16 MWh	2142.74 MWh
Cost per year for losses	289.861 k€	628.536 k€

7.3 Present Worth

The present worth for both cases is shown in Fig. 7.3.1 for case A and in Fig. 7.3.2 for case B. For both figures, the medium CAPEX cost shown in the previous sub-chapters and a load factor of 0.7 is used. Similar to other large power system components the CAPEX is the largest part.

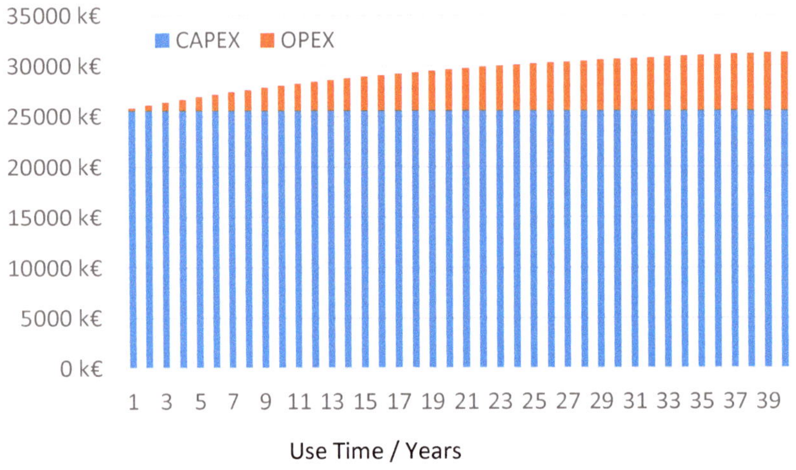

Figure 7.3.1: Present worth in € as a function of the use time of fault current limiter system for case A, load factor 0.7

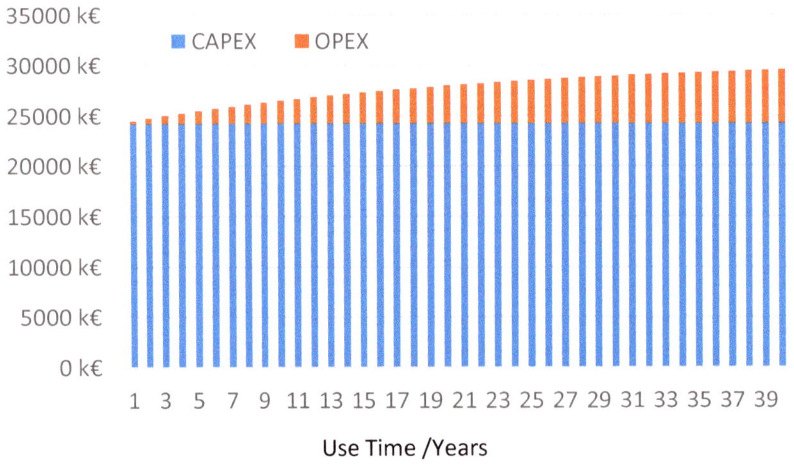

Figure 7.3.2: Present worth in € as a function of the use time of fault current limiter system for case B, load factor 0.7

8 Fault Analysis

8.1 Calculation of Symmetrical Faults at Short-Circuit Conditions

This chapter shows the current, temperature and resistance behaviour of a three phase limiter for symmetrical and unsymmetrical fault conditions. For unsymmetrical faults, a phase to phase fault and a phase to ground fault were considered. The simplified network equivalent is shown in Fig. 8.1.1. As can be seen, a constant voltage and a solid ground of the neutral point was assumed, due to a lack of more detailed data. The network impedance is represented by the impedance Z_{Line} and the fault location is between the impedance of the line and the load. The line impedance is adjusted to reach a continuous short-circuit current of 63 kA and the fault duration is 250 ms.

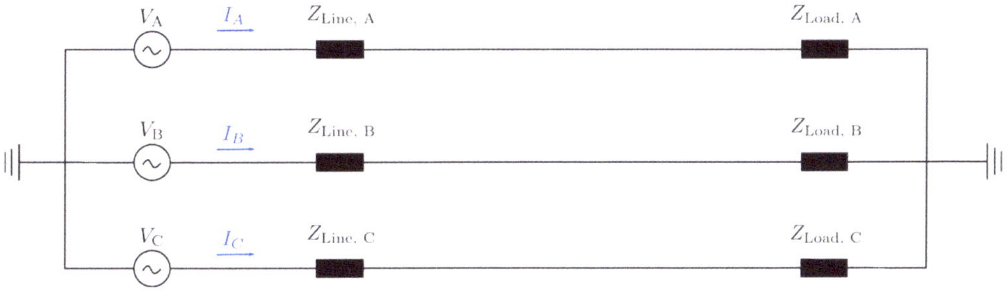

Figure 8.1.1: Simplified electrical network for the modelling of three phase current limitation

The current, temperature and resistance curves are shown in Fig. 8.2.1 for a symmetrical three-phase fault for case B. In the first half wave the superconductor in each phase becomes normal conducting when the current exceeds the critical current. Therefore, this time varies according to the current in each phase. Nevertheless, the maximum temperature and resistance at the end of the fault duration are very similar in each phase. In all phases, the current is limited below the specified value and does not exceed the maximum temperature of the superconductor.

8.2 Calculation of Unsymmetrical Faults at Short-Circuit Conditions

The current, temperature and resistance curves for a phase-to-phase fault are shown in Fig. 8.2.1 and a phase to ground fault in Fig. 8.2.2. It can be seen, that only the phase affected by the short-circuit limits the fault current. In both cases, the specification is fulfilled. For a more detailed investigation more information is needed on the specific short-circuit and grounding situation in the network.

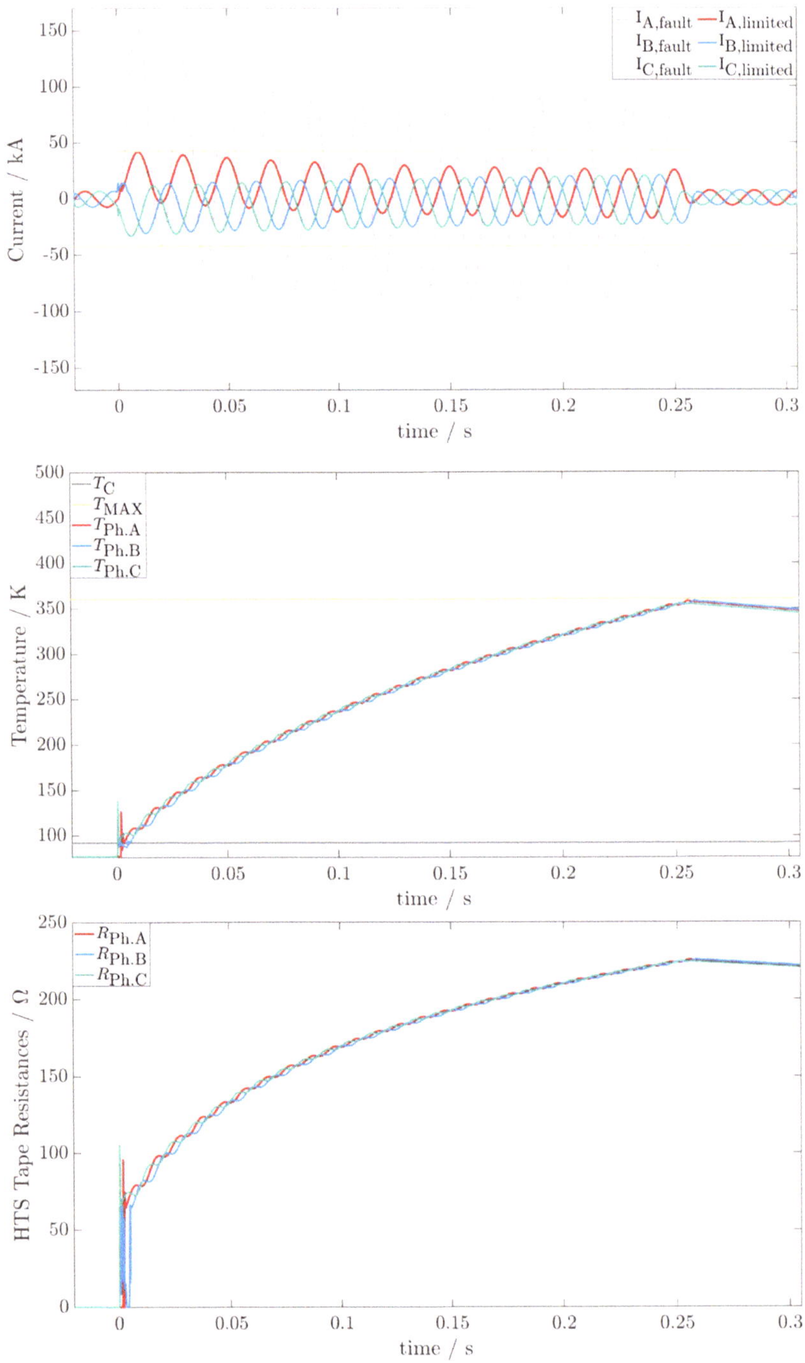

Figure 8.2.1: Current (top), temperature (middle) and resistance (bottom) curves for a three phase symmetrical fault for case B

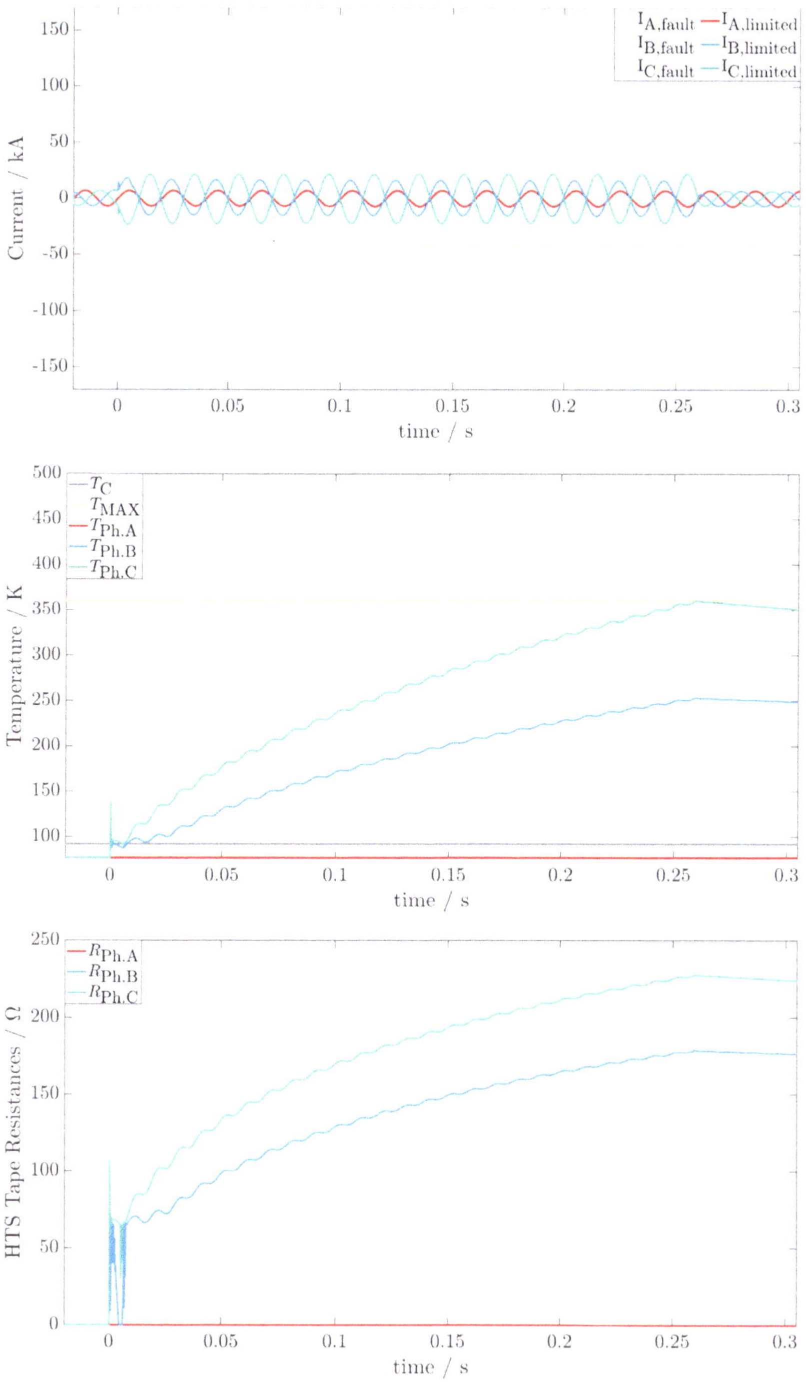

Figure 8.2.2: Current (top), temperature (middle) and resistance (bottom) curves for a fault between phase B and C for case B

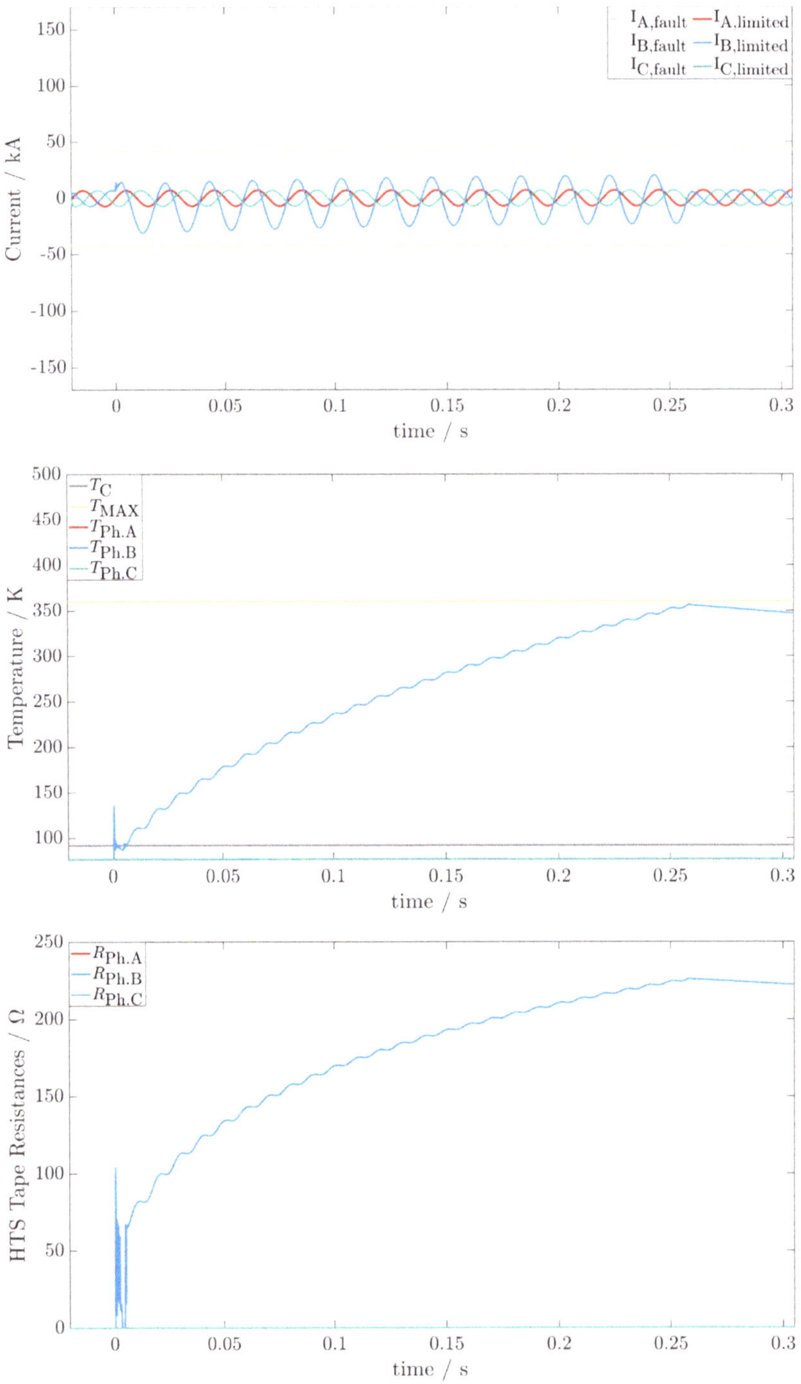

Figure 8.2.3: Current (top), temperature (middle) and resistance (bottom) curves for a phase B to ground fault for case B

9 State of the Art of Superconducting Fault Current Limiters

9.1 History and Medium Voltage SFCLs

Superconducting fault current limiters have been around for a long time, even before the discovery of high-temperature superconductivity in 1986. One example was a DC biased iron core SFCL with 3 kV and 556 A using NbTi superconductors [Raju1982]. Due to the large size and the high cooling effort, this activity was not continued. At this time, Alcatel Alsthom developed the first resistive type SFCLs with low temperature superconductors. The largest demonstrator with 36 kV, 210 A showed very good current limiting behaviour [Verhaege1996]. The main disadvantage with low-temperature superconductors immersed in liquid helium at 4.2 K was the fast release of helium during quench and the very high cooling effort. Therefore, R&D activity with low-temperature superconductors was discontinued.

After the discovery of HTS in 1986 by Bednorz and Müller R&D activities towards SFCLs were intensified. The first field test of a superconducting fault current limiter took place in 1996 in Switzerland. ABB developed a 10 kV, 70 A demonstrator and tested it for nearly one year in the hydro power station Lötzsch in Switzerland [Paul1997]. It was a so-called shielded-core-type fault current limiter that used Bi2212 bulk cylinders as the current limiting element. The current limiter shown in Fig. 9.1.1 was cooled with a liquid nitrogen dewar and had a consumption of 80 liters per day. During the one year test phase no short-circuit was observed in the grid and therefore no limitation was shown in the grid. ABB developed further with this material a 6.4 MVA resistive type SFCL demonstrator [Chen2002] but then stopped the R&D activities before further field tests.

Figure 9.1.1: 10 kV, 70 A shielded core type superconducting fault current limiter from ABB [Paul1997]

The first field test of a resistive type superconducting fault current limiter took place for one year in the RWE grid in Netphen in 2004 with a 10 kV, 600 A prototype. This limiter was developed from 2000 until 2003 in the German project CURL10 [Kreutz2005]. As shown in Fig. 9.1.2, the current limiting element was composed of Bi2212 bulk material components. In total 90 elements, 30 per phase, were installed in one cryostat with a maximum pressure of 10 bar. The limiter operated at 77 K in liquid nitrogen at ambient pressure conditions. The installation was in a standard container and current and voltage tests were performed before the grid connection. During the one year of installation time no three phase short-circuit occurred. Only single-phase short-circuits with lower short-circuit current occurred. Afterwards, Nexans further developed several medium voltage, resistive type superconducting fault current limiters with Bi2212 bulk material and performed several field tests. Due to the commercial availability of 2G coated conductor tapes from 2006 onwards and their improved flexibility and performance, the development of Bi2212 bulk material for fault current limiters was stopped.

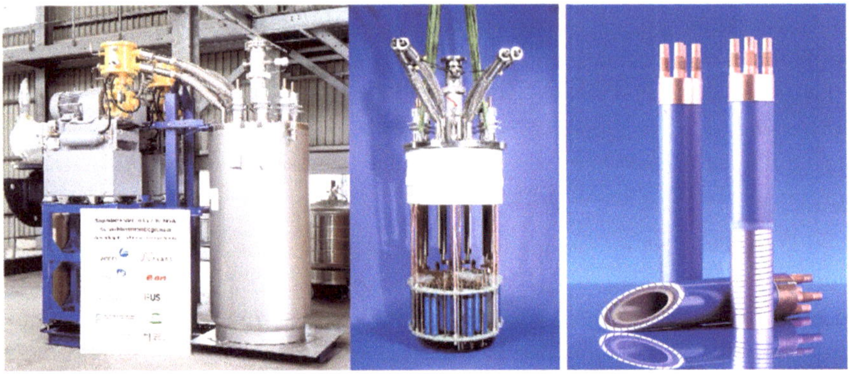

Figure 9.1.2: 10 kV, 600 A CURL10 fault current limiter (left) and Bi2212 bulk material current limiting element (right) [Bock2005, Kreutz2005]

One of the first resistive type superconducting fault current limiters with a field test using 2G coated conductor tapes was developed within the German Ensystrob project. The current limiting element was developed by Nexans and Forschungszentrum Karlsruhe [Elschner2012] and is shown in Fig. 9.1.3.

Figure 9.1.3: 2G coated conductor fault current limiting element (Photo: KIT)

This leads to further field tests with e.g. a 10 kV, 40 MVA limiter in Essen and a 12 kV, 1600 A limiter in Birmingham developed by Nexans. The limiter in Essen was developed within the German AmpaCity project [Stemmle2015] and operated continuously from 2014 until 2021 within the grid in Essen. With these limiters it was demonstrated that they can fulfil all system operator requirements and that a reliable operation is feasible. The limiter in Essen (see Fig. 9.1.4) is so far the one with the longest continuous operation time within a public grid.

Figure 9.1.4: 10 kV, 40 MVA resistive type superconducting fault current limiter [Schmidt2015]

In addition, Siemens and a few other companies further developed resistive type superconducting fault current limiters and tested them in the grid. One example is a 12 kV, 800 A prototype developed by Siemens and operated in Augsburg for several years [Kraemer2021].

It can be concluded that, at present, resistive type superconducting fault current limiters for medium voltage applications are technically mature. Several companies have developed first prototypes and operated them reliably in the grid for several years. With Nexans in France and LS in Korea, two companies exist that are able to develop and provide resistive type medium voltage superconducting fault current limiters.

9.2 High-Voltage SFCLs

In the last 10 years, a few resistive type superconducting fault current limiters were developed and realized worldwide for the AC transmission voltage level. The highest voltage reached so far is 220 kV for a limiter that is in operation in the grid in Moscow, Russia. The most important projects are described in the following, based on the published state of the art.

9.2.1 154 kV, 2 kA SFCL in South Korea

A 154 kV single-phase SFCL was developed in the years from 2011 to 2016 by the Korea Electric Power Corporation (KEPCO) Research Institute under the support of the Korean Ministry of Knowledge Economy. The single-phase system installed 2016 at the Gochang power test center of KEPCO consists of an HTS element, a current limiting reactor (CLR) and a circuit breaker (see Fig. 9.2.1). [Tixador2018, Lee2019].

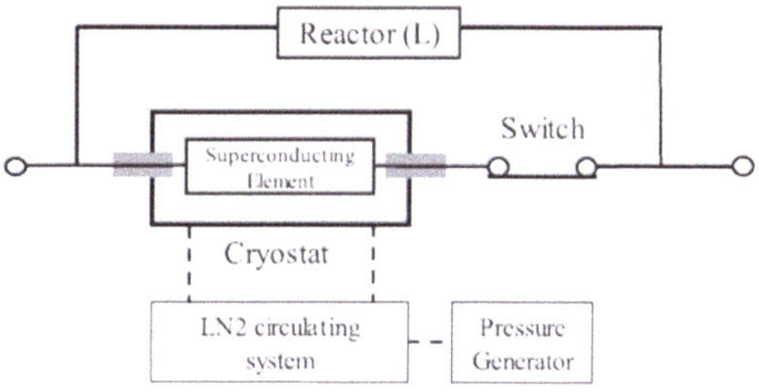

Figure 9.2.1: 154 kV single-phase SFCL system in South Korea [Lee2019, Han2019]

Table 9.2.1: Properties of 154 kV single-phase SFCL system in South Korea [Tixador2018, Lee2019, Han2019]

Nominal voltage	154 kV
Highest voltage for equipment	170 kV
AC withstand voltage (1 min)	325 kV
Lightning impulse voltage	750 kV
Nominal current	2 kA
Short-circuit current w/o limiter (rms)	52 kA
Limited short-circuit current (rms)	29 kA
Operating frequency	60 Hz
Fault duration	80 ms
Superconducting material	116 km of 2G YBCO
Impedance of current limiting reactor	1.52 Ω
Coolant	LN2 @ 5 bar, 71 K
Cooler	Stirling, 4 kW

The main technical properties of the system are summarized in Table 9.2.1. The system is designed for a nominal current of 2 kA and fulfils the requirements of equipment of 170 kV, which means an AC withstand voltage of 325 kV and a lightning impulse voltage of 750 kV. The complete system is able to limit a short-circuit current from 52 kA to 29 kA (rms). During current limiting and quench, the resistance of the HTS element increases from nearly zero to 9 Ω. For five cycles the current flows thought the superconducting element and the current limiting reactor until the circuit breaker opens and the current is only limited by the reactor with an impedance of 1.52 Ω. At an operating frequency of 60 Hz, this means a fault duration of 80 ms [Han2019].

The superconducting element consists of four modules in parallel and six modules in series, while in each module the HTS YBCO tape is wound in a meandering form to cancel out the magnetic field [Han2019]. In total, about 116 km HTS tape is used for the superconducting element [Lee2016].

Special attention was paid to the compactness of the cryostat of the superconducting element. A first model had an outer diameter of 3.1 m, an inner diameter of 2.7 m, a volume of 30.4 m^3 and a footprint of 17.7 m^2 and was designed with a maximum design field strength of 2.5 kV/mm at lightning impulse stress of 750 kV. This design considered a mixture of gas and liquid nitrogen above and partly lateral to the corona rings. Short-circuit tests showed that gas bubbles were only generated within the superconducting element surrounded by corona rings and disappeared before they reached the area outside of the corona rings. Therefore, the maximum design field strength was increased to 4.5 kV/mm based on the dielectric strength of liquid nitrogen and the second model of the cryostat was designed to be more compact with an outer diameter of 2.7 m, an inner diameter of 2.4 m, a volume of 19.7 m^3 and a footprint of 13.7 m^2. The electric field distribution and a cross-section of this compact model is shown in Fig. 9.2.2 [Han2019].

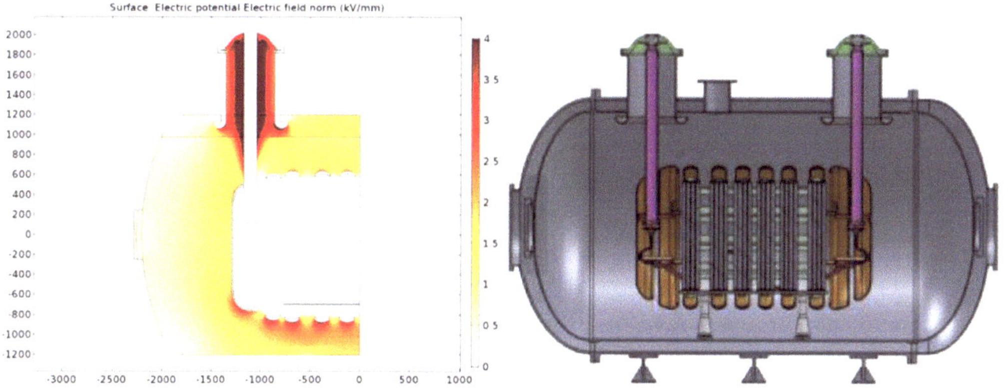

Figure 9.2.2: Electric field calculation at 750 kV lightning impulse voltage and cross-section of the second compact model of the cryostat with superconducting element [Han2019]

The compact model was tested successfully with positive and negative lightning impulse voltage of 750 kV (10 times each), AC short-time withstand voltage of 325 kV (1 min) and a long duration test according to IEC 60137 up to 170 kV. A four-month test with a line-to-ground voltage of 98 kV (= 154 kV/$\sqrt{3}$) was also successful [Han2019].

9.2.2 220 kV, 1.2 kA SFCL in Russia

The three-phase 220 kV SFCL system was developed between 2015 and 2019 and installed and finally tested September to December 2019 at the 220 kV "Mnevniki" substation in Moscow. The system was commissioned by the Moscow-owned electric utility company UNECO and built up by the company SuperOx. The SFCL system is installed parallel to an existing air-core reactor (ACR) in series with a 12 km long 220 kV cable (see Fig. 9.2.3). As far as is known, the complete system has been operating successfully since its installation [Moyzikh2021].

Table 9.2.2 summarizes the main properties of the three-phase system. The system is designed for a nominal current of 2 kA and a highest AC voltage for equipment of 252 kV, fulfilling an AC withstand voltage of 440 kV and a lightning impulse voltage of 950 kV. The system was tested to limit a short-circuit current in the first peak from 38 kA below 7 kA (peak). During current limiting and quench, the resistance of the superconducting element increases from nearly zero (0.01 Ω) to 40 Ω after 50 ms. The SFCL system is designed to withstand fault durations of 400 ms [Moyzikh2021].

The inner structure of the superconducting switching element is not published as far as known. Typical bifilar pancake coils are expected here. Corona rings to match the outer electric field strength to the cryostat inner wall (see Fig. 9.2.4 left) surround the switching element. The switching element has a length of 3.2 m and an outer diameter of 1.65 m, including the corona rings. The electric field strength at a lightning impulse voltage of 950 kV reaches values up to 4.8 kV/mm at the corona rings and about 9.7 kV/mm in the area of the bushings. These comparably high values result in a compact design of the cryostat with total outer dimensions of one phase of 5.5 m length, 2.85 m width and 6.5 m height including the bushings [Moyzikh2017].

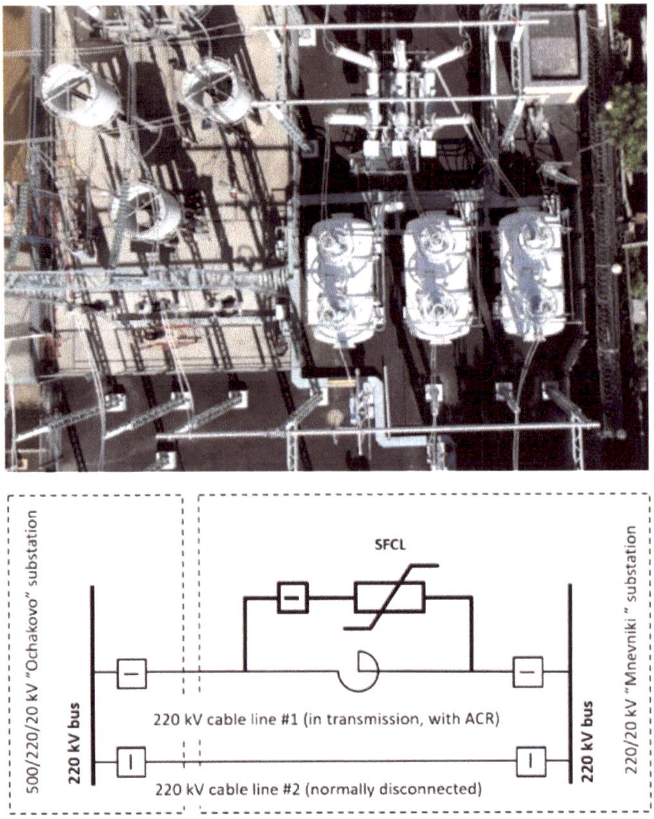

Figure 9.2.3: 220 kV three-phase SFCL system in Moscow [Moyzikh2021]

Table 9.2.2: Properties of 220 kV SFCL system in Moscow [Tixador2018, Moyzikh2021, Samoilenkov2020]

Nominal voltage	220 kV
Highest voltage for equipment	252 kV
AC withstand voltage (1 min)	440 kV
Lightning impulse voltage	950 kV
Switching impulse voltage*	325 kV
Nominal current	1.2 kA
Switching / critical current (peak)	3.4 kA
Short-circuit current w/o limiter (peak)	38 kA
Limited short-circuit current (peak)	6.8 kA
Operating frequency	50 Hz
Fault duration	400 ms
Superconducting material	3 x 8.4 km of 2G HTS (12 mm)
Impedance of air core reactor	3 Ω
Coolant	LN2 @ 5 bar, 65 – 77 K
Cryocooler	Brayton, 3 x 2 kW at 70 K

*) only for the bushings

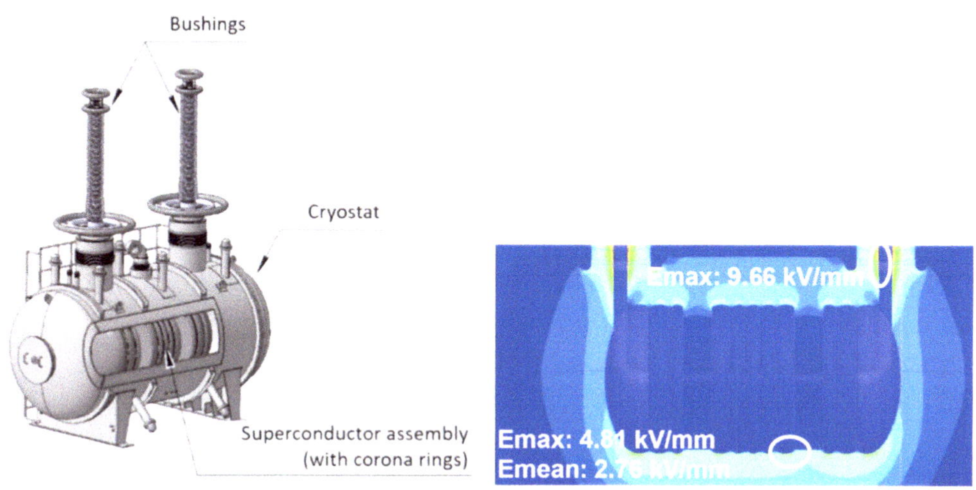

Figure 9.2.4: Scheme of one phase of the 220 kV SFCL and electric field calculation at 950 kV lightning impulse voltage [Moyzikh2021, Moyzykh2017, Tixador2018]

The Russian company Izolyator developed and manufactured the solid-state bushings. Each phase of the SFCL is equipped with its own cooling system (see Fig. 9.2.5) consisting of a pressure builder, a Turbo-Brayton NeoKelvin cryocooler from Taiyo Nippon Sanso and a circulation pump from Cryozone. The three systems are interconnecting, ensuring redundancy of the cooling system. [Moyzikh2021]

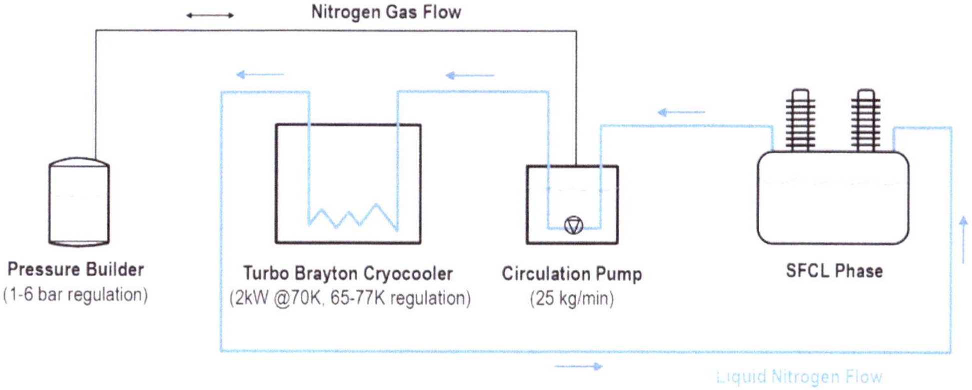

Figure 9.2.5: Cryogenic system of one phase of the 220 kV SFCL [Samoilenkov2020]

9.2.3 220 kV, 1.5 kA SFCL in China

A 220 kV SCFL was designed and developed in a cooperation between the Jiangsu Zhongtian Technology Company and the Beijing Jiaotong University. The design was first tested with a half-size module and several models of the cryogenic insulation supports to fulfil the requirements listed in Table 9.2.3. Later, one phase of the SFCL was built and tested up to an AC withstand voltage of 360 kV (1 min) and a lightning impulse voltage of 850 kV. Several short-circuit tests with prospective currents between 10 and 63 kA (rms) were performed with a current limiting reactor in parallel. [Zhang2019, Dai2019]

Table 9.2.3: Properties of the single-phase 220 kV SFCL system in China [Zhang2019, Dai2019]

Nominal voltage	220 kV
AC withstand voltage (1 min)	360 kV
Lightning impulse voltage	850 kV
Nominal current	1.5 kA
Switching / critical current	2.4 kA
Short-circuit current w/o limiter	63 kA
Limited short-circuit current	38 kA
Operating frequency	50 Hz
Fault duration	100 ms
Superconducting material	10.8 km of 2G YBCO (12 mm)
Coolant	LN_2 @ 5 bar

The superconducting element consists of 128 bifilar coils with an outer diameter of 788 mm and an inner diameter of 200 mm. 16 coils each are connected in parallel in a submodule and 8 of these submodules are connected in series resulting in a total length of 8.35 m of the superconducting element. The design of the SCFL is shown in detail in Fig. 9.2.6. The cryostat has inner and outer lengths of 12.9 m and 13.5 m, respectively, and inner and outer diameters of 2.5 m and 2.85 m, respectively, resulting in a total volume of 75.8 m³. This means that the SFCL is significantly larger compared to the limiter in Moscow at comparable voltage requirements [Zhang2019].

Figure 9.2.6: Sub-module of 16 parallel bifilar coils, assembly of a sub-modules within the cryostat, and SFCL under test [Zhang2019, Dai2019]

10 Summary and Conclusion

This study investigates the technical and economical feasibility of a resistive type 380 kV, 5 kA SFCL for two different specifications. Since, at present, SFCL do not exist for this very high voltage level emphasis was given to various high voltage design aspects.

Firstly, a detailed specification for a bus coupling (Case A) and a bus section coupling location (case B) was given in chapter 2. The main difference between these two scenarios is the maximum limited current with SFCL, which is lower in the bus coupling location with 19 kA in comparison to 30 kA in the bus section coupling.

In chapter 3 an overview of the superconducting material and manufacturers is given. It can be summarized that flat coated conductor tapes based on rare-earth Barium Copper Oxide should be preferred and that at least five companies can be considered for the delivery of the HTS material. The HTS material quantity needed for a 380 kV, 5 kA SFCL should be available within a reasonable project time. All manufacturers listed in this study claimed they can deliver the specified HTS material within less than a year.

Within chapter 4, the conceptual design, the amount of HTS material needed for a three-phase limiter was first calculated. Two basic configurations with and without a current limiting reactor in parallel to the SFCL were considered, with a clear preference for the scenarios with a current limiting reactor. This increases the reliability and reduces the amount of HTS material by approximately 10%. Based on the data of available HTS material, in total an HTS material amount of 10-12 mm wide tapes of 235 km for the bus coupling and 210 km for the bus section coupling would be needed, with a critical current of 700 A per tape at 77 K in self field conditions.

The HTS material has to be arranged in bifilar coils with 6 twin tapes in parallel to enable a compact and low AC loss design. In total 85 HTS current limiting modules with an outer diameter of 1100 mm is the preferred design option for the bus coupling and 93 HTS modules with an outer diameter of 1000 mm for the bus section coupling.

For a reliable high voltage design, a maximum electric field of 3.8 kV/mm within liquid nitrogen at 77 K and a pressure of 5 bar was assumed. This means, that the liquid nitrogen is operated in so-called sub-cooled conditions. For the critical areas within the limiter, design options were developed that keep the maximum electric field within this limit. Since the number and dimensions of the HTS current limiting modules for both specifications are very similar, one cryostat is proposed that fits for both specifications. The cryostat for a single phase of the SFCL has an outer diameter of 3.6 m, a total length of 13.6 m and a weight of 142 tons when filled with liquid nitrogen. The 380 kV bushing that can be operated at low temperatures within the cryostat has a total length of roughly 7 m, with its minimum flashover distance of 4354 mm in air.

The capacitive coupling of the current limiting coils can lead to a non-linear voltage distribution also during the limitation. Therefore, a detailed RLC network model was developed and the maximum voltage increase for one coil in comparison to an equal voltage distribution was calculated to be 131 % for the bus coupling and 140 % for the bus section coupling. For an insulation concept with Kapton within the limiting coil, the maximum voltages within the limiting module seem manageable.

Very fast transient voltages can be considered as travelling waves and the voltage distribution within a bifilar coil arrangement can be highly unsymmetrical. A detailed simulation model was developed and the results were compared with measurements carried out at a bifilar coil setup. The results show clearly that transient impulses propagate in radial and circumferential ways along the bifilar coil. Both parts have to be considered and lead to an increased voltage stress over the module and thus the winding insulation. The maximum voltage on the winding insulation during an LIC pulse was calculated to be roughly 160 kV at an oscillating voltage of about 8 MHz. In comparison with a maximum voltage of 24 kV for an RLC network this is much higher. Since the voltage stress decreases with increasing module number and in each module from winding to winding this indicates that only the modules at the beginning and the end of the SFCL and the outer windings in particular would require a reinforced winding insulation.

The complete SFCL system would require a volume of 16.66 m in length, 13.66 m in width and a height of 10.1 m for the three-phase SFCL arrangement. In addition, the current limiting reactors would require a volume of 10.3 m times 9.5 m with a height of 8.3 m. The cooling system can be accommodated in two standard containers and a separate storage tank.

The heat input to cryogenic temperatures was estimated for the various contributions and the maximum was 17.6 kW for the bus coupling in case A and 15.9 kW for the bus section coupling in case B. This value depends very much on the current and reduces to less than 2 kW with only 50 % current. In principle, this cooling capacity can be provided with an open or closed LN_2 cooling circuit and a detailed cooling scheme that would allow both options was proposed by industry. A similar cooling system has already been operated reliably for several years.

In chapter 5, internal and external failures and their consequences and mitigation measures were discussed. The failures were classified into three categories. Only internal short-circuits and extreme weather conditions were identified in the most severe category that would lead to a complete loss of the SFCL. This is similar to other equipment that operates in electric power systems.

At present, no IEC test standard exists for SFCL and therefore a test sequence was developed and proposed in chapter 6. This recommendation is based on the IEEE test guide "IEEE Guide for Fault Current Limiter (FCL) Testing of FCLs Rated above 1000 V AC" [IEEE2015] that covers most types of fault current limitation systems. Finally, an overview on possible test laboratories is given in this chapter.

In chapter 7, the economical feasibility of the SFCL is investigated with the present worth method. In detail the OPEX and CAPEX is shown and the cost of the main components is based on direct feedback and cost estimations from industry. Due to the preliminary nature of the study it was decided not to get full offers for the specific components. A cost range was given with minimum and maximum values for the most important components. With this approach, a medium investment cost of 25.8 Mio. € was estimated for the SFCL in bus coupling and 24.5 Mio. € for the bus section coupling location. The major cost part is still the HTS material with more than 40 % cost share. For the operation costs, approximately 300 k€ per year for both locations need to be considered for a load factor of 0.7.

Chapter 9 summarizes the state of the art of SFCL R&D. The first resistive type SFCL operated in a grid in Germany in 2004. Since then, several companies including Nexans and Siemens have several years of operation experience of resistive type SFCL for the medium voltage level. At present, no 380 kV resistive type SFCL exists but for a voltage of 220 kV first prototypes have been operated in the public power system.

It can be summarized that a 380 kV, 5 kA resistive type SFCL is technically feasible and that initially a dedicated R&D program is recommended for further development of the key critical components. It seems possible to develop such a limiter within the next five years. As an outcome of a first pilot model installation, more detailed economical evaluations must follow to allow a TOTEX comparison with conventional alternatives.

11 References

[ABB2017] ABB Broschüre Fault Current Limiters IS-limiter™ and FC-Protector® - The world's fastest switching devices Summary: The world's fastest switching devices Brochure - English - 2017-11-14 (https://new.abb.com/medium-voltage/apparatus/fault-current-limiters/current-limiter)

[Bauer2016] C. Bauer, et. al., Investigation of the Flashover and Breakdown Voltage of Thin Polymer Films in Liquid Nitrogen, IEEE Transactions on Applied Superconductivity, Vol. 26, No. 3, April 2016, doi: 10.1109/TASC.2016.2530863

[Bock2005] J. Bock, F. Breuer, H. Walter, S. Elschner, M. Klaimaier, R. Kreutz, M. Noe, CURL 10: Development and Field-Test of a 10 kV / 10 MVA resistive Current Limiter Based on Bulk MCP-BSCCO 2212, IEEE Transact. Appl. Supercond. 15 (2), 1955, 2005

[Buyanov1975] Y.L. Buyanov, A.B. Fradkov, I.Y. Shebalin, A review of current leads for cryogenic devices, Cryogenics, April 1975

[Chen2002] M. Chen, W. Paul, M. Lakner, L. Donzel, M Hoidis, P. Unternährer, R. Weder, M. Mendik, 6.4 MVA resitive fault current limiter based on Bi-2212 superconductor, Physica C 372–376 (2002) 1657–1663

[CIGRE2012] CIGRE Technical Brochure Nr. 497, Application and feasibility of fault current limiters in power systems, Working group A3-23, 2012

[CIGRE2016] CIGRE Technical Brochure, Common characteristics and emerging test techniques for high temperature superconducting power equipment, Working group D1.38, 2016

[CIGRE2021] CIGRE Technical Brochure, Electrical insulation systems at cryogenic temperatures, Working Group D1.64, 2021

[Clem2008] J. Clem, Field and current distributions and ac losses in a bifilar stack of superconducting strips, Physical Review, B 77, 134506, doi: 10.1103/PhysRevB.77.134506, 2008

[CISPR2017] CISPR TR 18-2, Radio interference characteristics of overhead power lines and high-voltage equipment - Part 2: Methods of measurement and procedure for determining limits, Edition 3.0 2017-10

[Dai2019] Shaotao Dai, Tao Ma, Chi Xue, Lianqi Zhao, Yong Huang, Lei Hu, Bangzhu Wang, Teng Zhang, Xiaofei Xu, Lingling Cai, Hao Chen, Development and test of a 220 kV/1.5 kA resistive type superconducting fault current limiter, Physica C: Superconductivity and its Applications, Volume 565, 2019, 1253501, ISSN 0921-4534, doi: 10.1016/j.physc.2019.06.004

[Dakin1970] T.W. Dakin, W. Luxa, G. Oppermann, J. Vigreux, G. Wind, H. Winkelnkemper, Breakdown of gases in uniform fields – Paschen curves for nitrogen, air and sulfur hexafluoride, ELECTRA Nr. 32, 1970

[Ekin2006] J. W. Ekin, Experimental techniques for low-temperature measurements: cryostat design, material properties, and superconductor critical-current testing, Oxford University Press, Oxford, United Kingdom, 2006, https://wiki.physics.wisc.edu/ObsCos/images/e/ea/Ekin_ExperimentalTechniquesForLowTemperatureMeasurements.pdf

[Elschner2012] S. Elschner, A. Kudymow, J. Brand, S. Fink, W. Goldacker, F. Grilli, M. Noe, M. Vojenciak, A. Hobl, M. Bludau, C. Jänke, S. Krämer, J. Bock, ENSYSTROB-Design, manufacturing and test of a 3-phase resistive fault current limiter based on coated conductors for medium voltage application, Physica C, 482, 98, 2012

[Fink1998] S. Fink, K. Bauer, D. Huber, A. Ulbricht, V. Zwecker, Entwurf des HTSL-Energiekabelteststandes HAIhKE für das Kryogene Hochspannungslabor, Forschungszentrum Karlsruhe (today: KIT), Interner Bericht F130.0030.0012/A, 1998

[Fink2014] S. Fink, H.-R. Kim, R. Müller, M. Noe, V. Zwecker, AC breakdown voltage of liquid nitrogen depending on gas bubbles and pressure, 2014 ICHVE International Conference on High Voltage Engineering and Application, doi: 10.1109/ICHVE.2014.7035445

[Gerhold2002] J. Gerhold, Cryogenic Liquids-A prospective insulation basis for future power equipment, IEEE Transactions on Dielectrics and Electrical Insulation, Vol. 9, No. 1, February 2002

[Gromoll2020] D. Gromoll, et. al., Dielectric strength of insulating material in LN2 with thermally induced bubbles, 14th European Conference on Applied Superconductivity (EUCAS 2019), doi: 10.1088/1742-6596/1559/1/012087

[Grover2018] F. W. Grover, Tables for the Calculation of the Inductance of Circular Coils of Rectangular Cross Section Volume Scientific Papers of the Bureau of Standards, Forgotten Books London April 2018, ISBN: 978-1333936259

[Han2019] Y.-H. Han et al., Development and Long-Term Test of a Compact 154-kV SFCL, IEEE Transactions on Applied Superconductivity, vol. 29, no. 4, pp. 1-6, June 2019, Art no. 5600106, doi: 10.1109/TASC.2018.2880325

[Hara1989] M. Hara, D.J. Kwak, M. Kubuki, Thermal bubble breakdown characteristics of LN2 at 0.1 MPa under a.c and impulse electric fields, Cryogenics 1989, Vol. 29

[HSP2021] 26 Pub. SETFt 24 – 800 kV, 06/12-1500 TH 102-120321, HSP Köln

[Hobl2013] Hobl, A.; Goldacker, W.; Dutoit, B.; Martini, L.; Petermann, A.; Tixador, P. (2013): Design and Production of the ECCOFLOW Resistive Fault Current Limiter. In: IEEE Trans. Appl. Supercond 23 (3), S. 5601804. DOI: 10.1109/TASC.2013.2238288

[IEEE2015] IEEE Guide for Fault Current Limiter (FCL) Testing of FCLs Rated above 1000 V AC, Switchgear Committee, IEEE Power and Energy Society, IEEE Std C37.302™-2015

[Koo2017] J. Koo, W. Shin, D. Oh, R. Hwang and B. Lee, Comparison of Surface Flashover Characteristics of Rod and Rib Type Post Insulator for Extra-High Voltage Superconducting Fault Current Limiter, Trans. on Appl. Supercond., vol. 27, no. 4, pp. 1-5, 2017, Art no. 7700805

[Kraemer2007] H.-P. Kraemer, et.al., Patent DE102006032702B3 Resistive supraleitende Strombegrenzeinrichtung mit bifilarer Spulenwicklung aus HTS-Bandleitern und Windungsabstandshalter, 2007-10-04, SIEMENS AG, Früheste Priorität: 2006-07-14, Früheste Veröffentlichung: 2007-10-04

[Kraemer2021] H.P. Krämer, et.al., ASSiST - a Superconducting Fault Current Limiter in a Public Electric Power Grid, IEEE Transactions on Power Delivery 2021, doi 10.1109/TPWRD.2021.3066845

[Kreutz2005] R. Kreutz et.al., System Technology and Test of CURL 10, a 10 kV, 10 MVA Resistive High-Tc Superconducting Fault Current Limiter, IEEE Trans. On Appl. Supercond., Vol. 15, No. 2, June 2005

[Kreuz2022] T. Kreuz, et. al., Lightning and Switching Impulse Strength of Insulating Materials in Tape Arrangements within Liquid Nitrogen, IEEE Transactions on Applied Superconductivity, Vol. 32, No. 4, June 2022, doi: 10.1109/TASC.2021.3136134

[Küchler2018] A. Küchler, High Voltage Engineering, 2018, Springer Vieweg Verlag, doi: 10.1007/978-3-642-11993-4

[Lee2016] H. Lee and H. Kang, Conceptual Design of a Resistive 154-kV/2-kA Superconducting Fault Current Limiter, IEEE Transactions on Applied Superconductivity, vol. 26, no. 4, pp. 1-5, June 2016, Art no. 5601805, doi: 10.1109/TASC.2016.2540652

[Lee2017] S. R. Lee, J.-J. Lee, J. Yoon, Y.-W. Kang and J. Hur, Protection Scheme of a 154-kV SFCL Test Transmission Line at the KEPCO Power Testing Center, IEEE Transactions on Applied Superconductivity, vol. 27, no. 4, June 2017, Art no. 5401305, doi: 10.1109/TASC.2017.2669159

[Lee2019] S. R. Lee, E. Y. Ko, J.-J. Lee and M.-C. Dinh, Development and HIL Testing of a Protection System for the Application of 154-kV SFCL in South Korea, IEEE Transactions on Applied Superconductivity, vol. 29, no. 5, Aug. 2019, Art no. 5602104, doi: 10.1109/TASC.2019.2897838

[Moyzykh2017] M. Moyzykh, Superconducting Fault Current Limiter for Moscow 220 kV City Grid, presentation at the 13th European Conference on Applied Superconductivity (EUCAS), 2017, Geneva, https://indico.cern.ch/event/659554/contributions/2695171/attachments/1525283/2384766/1LO1-02_Mikhail_Moyzykh_Room_1.pdf

[Moyzikh2021] M. Moyzikh, et. al., First Russian 220 kV Superconducting Fault Current Limiter (SFCL) For Application in City Grid, IEEE Transactions on Applied Superconductivity, Vol. 31, No. 5, August 2021, doi: 10.1109/TASC.2021.3066324

[NEMA2021] NEMA ST 20-2021, Dry-Type Transformers for General Applications, National Electrical Manufacturers Association-

[Noe2012] M. Noe, A. Hobl, P. Tixador, L. Martini and B. Dutoit, Conceptual Design of a 24 kV, 1 kA Resistive Superconducting Fault Current Limiter, IEEE Transactions on Applied Superconductivity, vol. 22, no. 3, June 2012, Art no. 5600304, doi: 10.1109/TASC.2011.2181284

[Noe2021] M. Noe, M. Bauer, Supraleitende Hochstromkabel und Leitungen: Ohne Widerstand in die Anwendung, Physik unserer Zeit 6/2021 (52), doi: 10.1002/piuz.202101615

[Norris1970] W.T. Norris, Calculation of hysteresis losses in hard superconductors carrying ac: isolated conductors and edges of thin sheets, Journal of Physics D: Applied Physics 3 489-507, doi: 10.1088/0022-3727/3/4/308

[Paul1997] W. Paul, M. Lakner, J. Rhyner, P. Unternahrer, Th. Baumann, M. Chen, L. Widenhorn and A. Guerig, Test of 1.2 MVA high-Tc superconducting fault current limiter, Supercond. Sci. Technol. 10 (1997) 914–918.

[Philippow1982] E. Philippow, Taschenbuch der Elektrotechnik, vol. 6, p. 351, Carl Hanser Verlag, 1982

[Radebaugh2004] R. Radebaugh, Refrigeration for Superconductors, Proceedings of the IEEE, Vol. 92, No. 10, October 2004, doi: 10.1109/JPROC.2004.833678

[Raju1982] B.P. Raju, T.C. Bartram; Fault current limiter with superconducting DC Bias, IEEE Proc. Vol. 129, No. 4, July 1982, pp.166

[Samoilenkov2020] S. Samoilenkov, 220 kV HTS-Strombegrenzer für das Stadtnetz in Moskau, presentation at VII. ZIEHL conference, 2020, Berlin, https://ivsupra.de/wp-content/uploads/2020/03/4_Samoilenkov-ZIEHL-VII-Netzausbau-mit-Supraleitung-I.pdf

[Schacherer2009] C. Schacherer, Theoretische und experimentelle Untersuchungen zur Entwicklung supraleitender resistiver Strombegrenzer, Dissertation, Karlsruhe: Universitätsverlag Karlsruhe, 2009

[Schmidt2014] W. Schmidt, Auslegung supraleitender Strombegrenzer im Hochspannungsnetz, presentation at IV. ZIEHL conference, March 2014, Bonn, https://ivsupra.de/iv-ziehl-vortraege/

[Schmidt2015] F. Schmidt, Advanced Superconducting 10 kV System in the City Center of Essen, Germany, EUCAS Conference 2015, Lyon, https://snf.ieeecsc.org/sites/ieeecsc.org/files/documents/snf/abstracts/revStemmle1A-LS-O1.1.pdf

[Stemmle2015] M. Stemmle, F. Merschel, M. Noe, Update on world's first superconducting cable and fault current limiter installation in a German city center, 23rd International Conference on Electricity Distribution Lyon, 15-18 June 2015

[Tixador2018] P. Tixador, Superconducting Fault Current Limiter: Innovation for the Electric Grids (World Scientific Series in Applications of Superconductivity and Related Phenomena, World Scientific Publishing Co Pte Ltd. November 2018, ISBN: 978-9813272972

[Tixador2019] P. Tixador et. al., Status of the European Union Project FASTGRID, IEEE Transactions on Applied Superconductivity, Vol. 29, Issue 5, Aug. 2019, doi: 10.1109/TASC.2019.2908586

[VDI2013] VDI Wärmeatlas 2013, 11. bearbeitete und erweiterte Auflage, Springer-Verlag Berlin Heidelberg 2013, ISBN 978-3-642-19980-6, doi: 10.1007/978-3-642-19981-3

[Verhaege1996] T. Verhaege et.al. Experiments with a high voltage (40 kV) superconducting fault current limiter, Cryogenics 36 (1996) 521-526

[Wiedemann-Frantz1853] G. Wiedemann, R. Franz: Ueber die Wärme-Leitungsfähigkeit der Metalle. In: Annalen der Physik. Band 165, Nr. 8, 1853, S. 497–531, doi: 10.1002/andp.18531650802

[Wilson1987] M. Wilson, Superconducting Magnets, Clarendon Press, 1987, ISBN-10:0198548109

[Zhang2019] Zhang, J., Teng, Y., Qiu, Q. et al., Design and Development of a 220-kV Resistive-Type Fault Current Limiters Based on 2G-Coated Conductors, J Supercond Nov Magn 32, 3779–3787, 2019, doi: 10.1007/s10948-019-05188-4

12 Annex

12.1 HTS Material Data

Product Type	TPL1100	TPL1120	TPL2100	TPL2120	TPL4121	TPL5121
Description	HTS wire with surround silver metallization	thin HTS wire with surround silver metallization	HTS wire with laminated Cu stabilization	thin HTS wire with laminated Cu stabilization	thin HTS wire with surround copper stabilization	thin HTS wire with surround copper stabilization and SnPb solder coating
Typical applications	stacks or stacked conductors for levitation and cables	high current density stacks or stacked conductors for levitation and cables	medium magnetic field coils for rotating machines etc.	medium magnetic field coils for rotating machines etc.	coils or stacks with very high current density, small bending radii designs	solder cast coils and stacked conductors
Substrate	100 µm Hastelloy C276	50 µm Hastelloy C276	100 µm Hastelloy C276	50 µm Hastelloy C276	50 µm Hastelloy C276	50 µm Hastelloy C276
Buffer layer	MgO	MgO	MgO	MgO	MgO	MgO
Superconducting layer	GdBaCuO	GdBaCuO	GdBaCuO	GdBaCuO	GdBaCuO	GdBaCuO
Metallization	Ag	Ag	Ag	Ag	Ag	Ag
Stabilization	none	none	100 µm Cu soldered on HTS side	100 µm Cu soldered on HTS side	10 µm Cu surround - different thicknesses available on request	10 µm Cu surround coated with a thin layer of SnPb40
Thickness	0.11 mm	0.06 mm	0.22 mm	0.17 mm	0.08 mm	0.08 mm
Width	12 mm	12 mm	12 mm	12 mm	12 mm	12 mm
Min. double bending diameter (RT, 95 %)	60 mm	40 mm	60 mm	40 mm	40 mm	40 mm
Recommended max. handling force	100 N (10 kg)	50 N (5 kg)	100 N (10 kg)	50 N (5 kg)	50 N (5 kg)	50 N (5 kg)
Maximum rated stress (RT)	600 MPa	600 MPa	300 MPa	300 MPa	500 MPa	500 MPa
Maximum rated tensile strain (95 %)	0.30 %	0.30 %	0.30 %	0.30 %	0.30 %	0.30 %
I_c (77 K, sf)	typical 360 A - 500 A, higher values available on request					
Piece length	typical 25 m - 200 m, longer pieces available on request					

August 2020
THEVA Dünnschichttechnik GmbH • Rote-Kreuz-Str. 8 • 85737 Ismaning • Germany • info@theva.com • www.THEVA.com

Figure 12.1.1: Theva TPL1100 material data

12.2 Material Data used for the Simulation of the HTS tapes

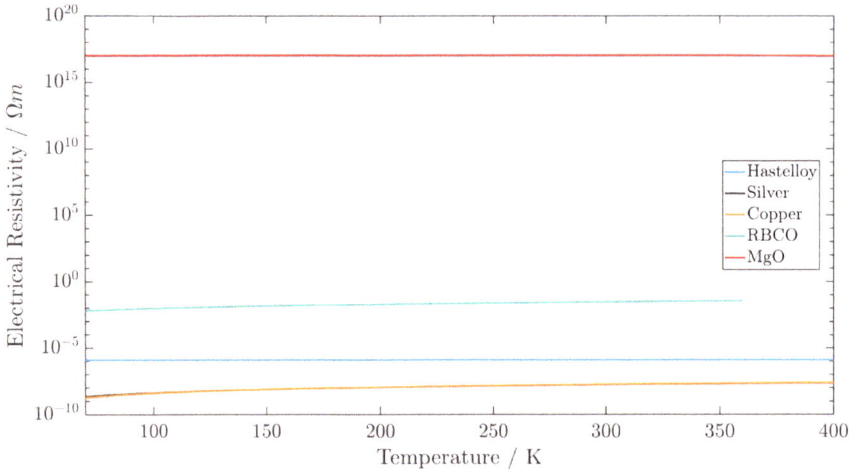

Figure 12.2.1: Temperature dependence of electrical resistivity of the materials within an HTS tape

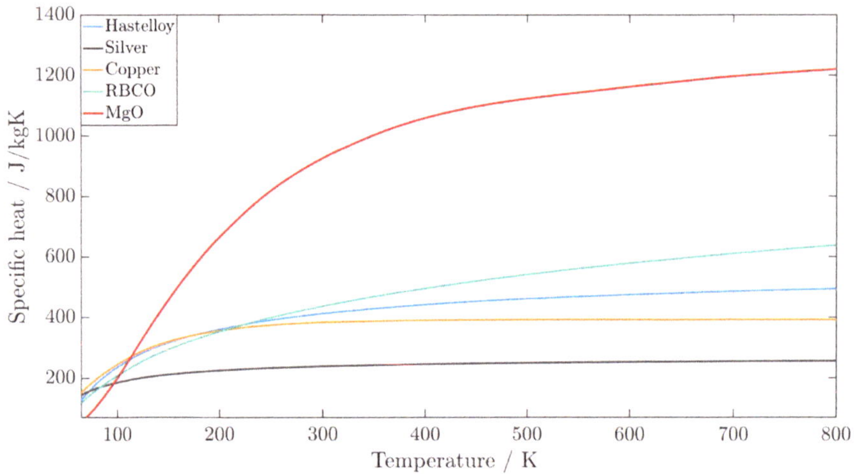

Figure 12.2.2: Temperature dependence of specific heat of the materials within an HTS tape

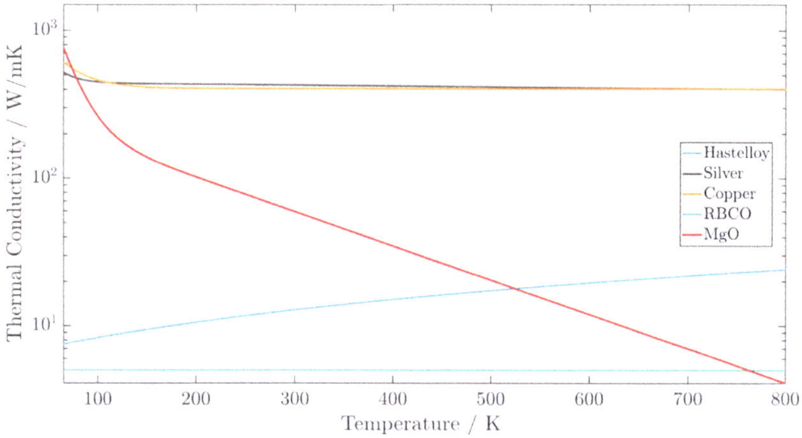

Figure 12.2.3: Temperature dependence of thermal conductivity of the materials within an HTS tape

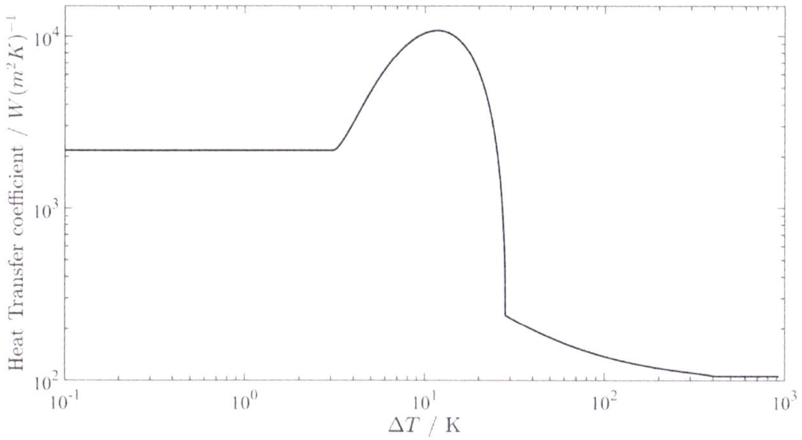

Figure 12.2.4: Heat transfer coefficient from solid material to liquid nitrogen at 77 K, 1 bar

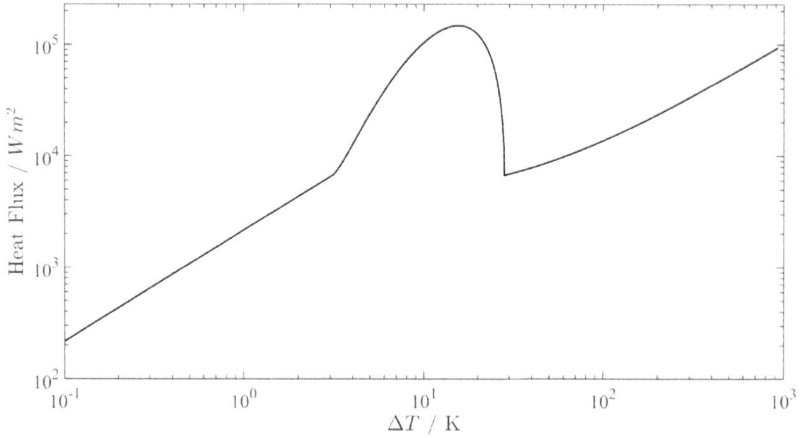

Figure 12.2.5: Heat flux from solid material to liquid nitrogen at 77 K, 1 bar

Karlsruher Schriftenreihe zur Supraleitung

Karlsruher Institut für Technologie (KIT) | ISSN 1869-1765

Band 001 **Christian Schacherer**
Theoretische und experimentelle Untersuchungen zur Entwicklung supraleitender resistiver Strombegrenzer.
ISBN 978-3-86644-412-6

Band 002 **Alexander Winkler**
Transient behaviour of ITER poloidal field coils.
ISBN 978-3-86644-595-6

Band 003 **André Berger**
Entwicklung supraleitender, strombegrenzender Transformatoren.
ISBN 978-3-86644-637-3

Band 004 **Christoph Kaiser**
High quality Nb/Al-AlOx/Nb Josephson junctions. Technological development and macroscopic quantum experiments.
ISBN 978-3-86644-651-9

Band 005 **Gerd Hammer**
Untersuchung der Eigenschaften von planaren Mikrowellenresonatoren für Kinetic-Inductance Detektoren bei 4,2 K.
ISBN 978-3-86644-715-8

Band 006 **Olaf Mäder**
Simulationen und Experimente zum Stabilitätsverhalten von HTSL-Bandleitern.
ISBN 978-3-86644-868-1

Band 007 **Christian Barth**
High Temperature Superconductor Cable Concepts for Fusion Magnets.
ISBN 978-3-7315-0065-0

Die Bände sind unter www.ksp.kit.edu als PDF frei verfügbar oder als Druckausgabe bestellbar.

Band 008 **Axel Stockhausen**
Optimization of Hot-Electron Bolometers for THz Radiation.
ISBN 978-3-7315-0066-7

Band 009 **Petra Thoma**
Ultra-fast $YBa_2Cu_3O_{7-x}$ direct detectors for the THz frequency range.
ISBN 978-3-7315-0070-4

Band 010 **Dagmar Henrich**
Influence of Material and Geometry on the Performance of Superconducting Nanowire Single-Photon Detectors.
ISBN 978-3-7315-0092-6

Band 011 **Alexander Scheuring**
Ultrabreitbandige Strahlungseinkopplung in THz-Detektoren.
ISBN 978-3-7315-0102-2

Band 012 **Markus Rösch**
Development of lumped element kinetic inductance detectors for mm-wave astronomy at the IRAM 30 m telescope.
ISBN 978-3-7315-0110-7

Band 013 **Johannes Maximilian Meckbach**
Superconducting Multilayer Technology for Josephson Devices.
ISBN 978-3-7315-0122-0

Band 014 **Enrico Rizzo**
Simulations for the optimization of High Temperature Superconductor current leads for nuclear fusion applications.
ISBN 978-3-7315-0132-9

Band 015 **Philipp Krüger**
Optimisation of hysteretic losses in high-temperature superconducting wires.
ISBN 978-3-7315-0185-5

Band 016 **Matthias Hofherr**
Real-time imaging systems for superconducting nanowire single-photon detector arrays.
ISBN 978-3-7315-0229-6

Band 017 **Oliver Näckel**
Development of an Air Coil Superconducting Fault Current Limiter.
ISBN 978-3-7315-0526-6

Band 018 **Christoph M. Bayer**
Characterization of High Temperature Superconductor Cables for Magnet Toroidal Field Coils of the DEMO Fusion Power Plant.
ISBN 978-3-7315-0605-8

Band 019 **Shengnan Zou**
Magnetization of High Temperature Superconducting Trapped-Field Magnets.
ISBN 978-3-7315-0715-4

Band 020 **Ilya Charaev**
Improving the Spectral Bandwidth of Superconducting Nanowire Single-Photon Detectors (SNSPDs).
ISBN 978-3-7315-0745-1

Band 021 **Juliane Raasch**
Electrical-field sensitive $YBa_2Cu_3O_{7-x}$ detectors for real-time monitoring of picosecond THz pulses.
ISBN 978-3-7315-0786-4

Band 022 **Yingzhen Liu**
Design of a superconducting DC wind generator.
ISBN 978-3-7315-0796-3

Band 023 **Sebastian Hellmann**
Research and Technology Development on Superconducting Current Limiting Transformers.
ISBN 978-3-7315-0804-5

Band 024 **Simon J. Otten**
Characterisation of REBCO Roebel cables.
ISBN 978-3-7315-0904-2

Band 025 **Julia Brandel**
Supraleitende Einzelphotonenzähler: Optimierung der Zeitauflösung und Anwendungsbeispiele aus der Spektroskopie.
ISBN 978-3-7315-0917-2

Band 026 **Dustin Kottonau, Eugen Shabagin, Wescley T. B. de Sousa, Jörn Geisbüsch, Mathias Noe, Hanno Stagge, Simon Fechner, Hannes Woiton, Thomas Küsters**
Bewertung des Einsatzes supraleitender 380-kV-Kabel.
ISBN 978-3-7315-0927-1

Band 027 **Steffen Dörner**
Multifrequenzausleseverfahren von supraleitenden Einzelphotonen-Detektoren.
ISBN 978-3-7315-0961-5

Band 028 **Michael Merker**
Superconducting integrated THz receiver.
ISBN 978-3-7315-0970-7

Band 029 **Wolfgang-Gustav Ekkehart Schmidt**
Superconducting Nanowire Single-Photon Detectors for Quantum Photonic Integrated Circuits on GaAs.
ISBN 978-3-7315-0980-6

Band 030 **Dustin Kottonau, Eugen Shabagin, Wescley de Sousa, Jörn Geisbüsch, Mathias Noe, Hanno Stagge, Simon Fechner, Hannes Woiton, Thomas Küsters**
Evaluation of the Use of Superconducting 380 kV Cable.
ISBN 978-3-7315-1026-0

Band 031 **Alan Preuß**
Development of high-temperature superconductor cables for high direct current applications.
ISBN 978-3-7315-1041-3

Die Bände sind unter www.ksp.kit.edu als PDF frei verfügbar oder als Druckausgabe bestellbar.

Band 032 **Roland Gyuráki**
Fluorescent thermal imaging method for investigating transient effects in high-temperature superconductor tapes and coils.
ISBN 978-3-7315-1064-2

Band 033 **Aurélien Godfrin**
AC Losses in High-Temperature Superconductor Tapes and Cables for Power Applications.
ISBN 978-3-7315-1096-3

Band 034 **Alexander Schmid**
Integrierte planare Antennen für supraleitende Detektorsysteme zur THz-Strahldiagnostik.
ISBN 978-3-7315-1145-8

Band 035 **Alexander Buchholz**
Prospective Life Cycle Assessment of High-Temperature Superconductors for Future Grid Applications.
ISBN 978-3-7315-1194-6

Band 036 **Emanuel Marius Knehr**
Technology and readout for scaling up superconducting nanowire single-photon detectors.
ISBN 978-3-7315-1256-1

Band 037 **Mathias Noe, Chistof Humpert, Wescley de Sousa, Stefan Fink, Friedhelm Herzog, Stefan Huwer, Marvin Nießen, Sebastian Pfaller, Stephan Pöhler, Tobias Rodler, André Schmid, Tabea Arndt**
380 kV Superconducting Fault Current Limiter Feasibility Study.
ISBN 978-3-7315-1313-1